ANALYTICAL MOLECULAR BIOLOGY

ANALYTICAL MOLECULAR BIOLOGY

by

Tai Te Wu
Northwestern University

SPRINGER SCIENCE+BUSINESS MEDIA, LLC

Library of Congress Cataloging-in-Publication

Analytical molecular biology / edited by Tai Te Wu.
p. cm.
ISBN 978-1-4757-7481-8 ISBN 978-0-306-46892-6 (eBook)
DOI 10.1007/978-0-306-46892-6
1. Molecular biology—Mathematics I. Wu, Tai Te, 1935-
AH 506.A535 2001
572.8'01'51--dc21 2001038129
CIP

Originally published by Kluwer Academic Publishers in 2001
Softcover reprint of the hardcover 1st edition 2001

This printing is a digital duplication of the original edition.

Printed on acid-free paper.

Dedicated to all my students
of the last forty years.

CONTENTS

PREFACE

This book is designed to illustrate the importance of analytical methods applied to some basic molecular biology problems. It serves two purposes: (1) for molecular biologists to learn some mathematics, and (2) for applied mathematicians to learn some molecular biology. As our starting point, some knowledge of both mathematics and molecular biology has to be assumed. We would like to begin at the level of high school advanced placement courses. In addition, it is important to try to solve fundamental molecular biology problems with simple applied mathematical methods, rather than looking for complicated mathematical equations in less important problems in biology.

To illustrate this approach, eight fundamental problems in molecular biology will be discussed in this book in the order of mathematical complexity. There are, of course, many more other important problems. However, after going over this book, the readers can develop their own analytical methods to study any problem of particular interest to them. The basic philosophy is that we are going to start with crucial experimental data for a specific problem in molecular biology. A simple model will then be constructed with the inclusion of as much biological facts as possible. From this model, explicit predictions can then be deduced and in turn suggest further experimental studies.

Chapter 1 discusses the important experiments on amino acid sequencing of Bence Jones proteins, their relationship to antibodies or immunoglobulins, alignment of these sequences, calculation of variability as function of position to locate antibody combing sites, structures of antibody-antigen complexes, humanization of rodent antibodies for therapeutic use, etc.

Chapter 2 considers thc saturation of hemoglobin with oxygen at equilibrium. This classical problem has been analyzed by many famous molecular biologists. A particular set of extremely accurate experimental measurements will be graphed in the Hill's plot. Three models, one-constant (Hill), two-constant (Paulin) and three-constant (MWC), are discussed in detail. Non-linear least square fitting of these theoretical equations to the experimental data in the Hill's plot gives the equilibrium constants indicative of allosteric transition. Further experiments provide a detailed molecular mechanism of this transition.

In order to sequence the entire genome of *Escherichia coli,* the locations of over one thousand genetic markers have been precisely positioned on its chromosomal map by numerous transduction experiments. In Chapter 3, a mathematical model is constructed based on all known biological processes involved in transduction, so that co-transduction frequencies can be converted into distances between genetic markers. This traditional map provides the basis for the eventual construction of the physical map.

Most of the metabolic pathways involve a collection of different enzymes. To simplify experimental studies, each enzymatic reaction is usually measured separately by mixing purified enzyme molecules with substrate and co-enzyme. Even in such isolated measurements, the analytical equations are still too complicated. In Chapter 4, simple enzyme kinetic equations are derived as a set of non-linear ordinary differential equations. Various steps of further simplification are illustrated in detail. One of the original sets of experimental measurements by Michaelis and Menten is then compared with these theoretical results.

For 40 years, many three dimensional structures of proteins have been determined by X-ray diffraction studies, and to a lesser extent by nuclear magnetic resonance studies. Atomic coordinates of non-hydrogen atoms are available from Protein Data Bank. Chapter 5 explains how these coordinates can be used to calculate bond distances, bond angles, ϕ and ψ angles of the protein backbone. Furthermore, hydrogen atoms can be theoretically positioned into these three dimensional structures, in order to analyze steric hindrance. Sterically hindered (ϕ,ψ) combinations should be avoided in future refinements.

In Chapter 6, one of the predictive methods of estimating protein backbone (ϕ,ψ) angles based on the local amino acid sequence and side chain χ angles based on steric hindrance is illustrated. This approach is then applied to predict three-dimensional structures of short peptide hormones, and of antibody complementarity determining regions. Other analytical methods of predicting protein tertiary structures based on their primary sequences should be developed, since many amino acid sequences become available from various genome projects.

The backbone structure of nucleic acids is much more complicated. Chapter 7 explains that each nucleotide unit has five single bonds with rotational degree of freedom and some flexibility of its ribose ring. Thus, a minimum of a five-dimensional space will be required for detailed analysis of its possible configurations without steric hindrance. Attempts to simply this complicated problem with two virtual bonds are discussed. Due to its complexity, analytic studies of three-dimensional structures of nucleotides and nucleic acids will be a very challenging problem for both molecular biologists and applied mathematicians.

In Chapter 8, the classical problem of DNA double helix is discussed. In order to analyze the X-ray diffraction pictures of DNA fibers at 92% and 66% relative humidity, we have to thoroughly understand some of the basic properties of Fourier transform, Bessel functions, complex variables, etc. It is hoped that this chapter can eliminate the lack of knowledge of these mathematical methods among molecular biologists. Therefore, in the Appendix, some of the basic tools of applied mathematics are summarized.

Finally, models are by definition simplified versions of the real molecular biology processes. Therefore, constant improvements are essential as more experimental data become available. Readers are encouraged to make such modifications, as well as to propose their own analytical approaches to the problems mentioned in this book and other important molecular biology problems.

CHAPTER 1

ANTIBODIES BINDING ANTIGENS

INTRODUCTION

Under normal circumstances, we are constantly exposed to foreign substances including chemicals, bacteria, viruses, etc. In order to defend ourselves against the adverse effects of these substances, we have developed various immunological mechanisms over years of evolution. The so-called innate process involves certain cells capable of eating such foreign substance. For example, carbon particles from smoke are picked up by macrophages lining our lungs. The macrophages cannot digest these carbon particles which in turn are not toxic. They just stay around. For city dwellers, their lungs gradually turn black after many years of such exposure. On the other hand, if the foreign substance is a pathogenic bacterium, the macrophage may try to digest the bacterium. Or, on entering the macrophage, the bacterium may survive and start to divide, thus killing the macrophage. Such immunological battles go on continuously during our entire life.

In order to improve our chance of survival, we have developed a more efficient line of defense. If our macrophages have managed to kill off a certain type of pathogenic bacteria, the so-called adaptive process of defense somehow remembers this incident, by producing special protein molecules known as antibodies. If we encounter the same bacteria a second time, one end of these antibodies will be able to bind to some surface molecules of these bacteria known as antigens. The other end of the antibodies will bind to certain special receptors on the surface of the macrophages to improve the efficiency of these macrophages to eat them. Since the number of different antigens in the world is of the order of millions or billions, we need to have a system capable of producing millions or billions of different antibodies just for survival in the existing environment. Any deficiency of this system due to genetic defect, infection, aging, etc. can result in disease processes. In fact, we are able to produce specific antibodies capable of binding foreign substances newly synthesized in the laboratory and have never appeared in our evolutionary past.

BEGINNING OF IMMUNOCHEMISTRY

It is thus extremely difficult for biochemists and molecular biologists to isolate one of these antibodies out of million or billion different ones in our blood, and try to characterize its properties. On the other hand, over one hundred years ago in 1848, a physician, Dr. Bence Jones, noticed patients suffering from a type of cancer, multiple myeloma, excreted a protein in their urine. On heating the urine in slight acid solution, he saw the protein precipitating out of solution at 56-64^0. It would re-dissolve on boiling (Kabat, 1976). That protein is now known as Bence Jones protein named after the discoverer. Subsequently, it was also found that such patient had large amounts of a specific myeloma protein in their blood, different from patient to patient. It appears as a large peak on starch gel electrophoresis of serum proteins, and can thus be purified. However, its function is not known. These proteins are known as immunoglobulins. A graduate student at Rockefeller University, Edelman (1959), discovered that on reducing the disulfide bonds the myeloma protein from a multiple myeloma patient consisted of two chains of polypeptides, one heavy and one light, based on their molecular weights. The light one corresponded to the Bence Jones protein in the urine of the same patient. In the same year, Porter (1959) showed that rabbit antibodies could be digested by papain to give two distinct fractions, one could still bind antigen while the other had no antibody activity. The combination of these two important discoveries provided biochemists and molecular biologists the handle to start the investigation of the properties of antibodies.

We now know that multiple myeloma is a cancer of one of the antibody producing cells, and immunoglobulin is a generalized term for antibodies. Indeed, the disease process, multiple myeloma, has achieved the task of isolating one of the million or billion antibodies from the rest. In 1965, Hilschmann and Craig at Rockefeller University determined the amino acid sequences of the first three Bence Jones proteins from three different patients and opened up a new scientific field known as immunochemistry. Even though the three proteins were not completely sequenced by them at the time of the publication of their important paper, these sequences subsequently became available as shown in Fig. 1-1. For conciseness, as commonly used in any biochemistry or molecular biology textbook, the single letter abbreviations for amino acid residues are used.

```
        1         10        20                30
        |         |         |                 |
Roy    -DIQMTQSPSSLSASVGDRVTITCQASQ------SISIFLNW
Ag     -DIQMTQSPSSLSASVGDRVTITCQASQ------SISIFLNW
Cum    EDIVMTQTPLSLPVTPGEPASISCKSSQSLLDSGDGNTYLNW

          40        50        60        70
          |         |         |         |
Roy    YQQKPGKAPKLLYIDASKLEAGVPSRFSGTGSGTDFTFTISG
Ag     YQQGPKKAPKLLYIDASNLETGVPSRFSGSGFGTDFTFTISG
Cum    YLQKAGNSPQLLIYTLSYRASGVPDRFSGSGSGTDFTLKISR

        80        90            100     107
        |         |             |       |
Roy    LQPEDIATYYCQQFDNLP------LTFGGGTKVPF-K
Ag     LQPEDIATYYCQQYDTLP------KTFGQGTKLEI-K
Cum    VQAEDVGVYYCMQRLEIP------YTFGQGTKLEI-K

       110       120       130       140
        |         |         |         |
Roy    RTVAAPVSFIFPPSDEQLKSGTASVVCLLNNFYPREAKVQWL
Ag     RTVAAPVSFIFPPSNEQLKSGTASVVCLLNNFYPREAKVQWL
Cum    RTVAAPVSFIFPPSDEQLKSGTASVVCLLNNFYPREAKVQWL

     150       160       170       180       190
       |         |         |         |         |
Roy    VDNALQSGNSQESVTQQDSKDSTYSLSSTLTLSKADYEKHKL
Ag     VDNALQSGNSQESVTEQDSKDSTYSLSSTLTLSKADYEKHKV
Cum    VDNALQSGNSQESVTQQDSKDSTYSLSSTLTLSKADYEKHKV

             200       210 214
              |         |   |
Roy    YACEVTHQGLSSPVTKSFNRYEC
Ag     YACEVTHQGLSSPVTKSFNRYEC
Cum    YACEVTHQGLSSPVTKSFNRYEC
```

Figure 1-1. Amino acid sequences of three Bence Jones proteins, Cum, Ag and Roy, aligned with the Kabat numbering system (Kabat *et al.*, 1991).

Each sequence consists of approximately 214 amino acid residues as listed from the N-terminal to the C-terminal. The sequences of the first 107 amino acid residues of the three proteins were different, and Hilschmann and Craig designed that portion of these proteins as the variable region. On the other hand, the remaining 107 amino acid residues, i.e. from position 108 to 214, had the same sequence except possibly two or three positions. Thus, that portion was known as the constant region. They made a crucial prediction, i.e. the variable and constant regions of the same protein were coded by two different genes.

To sequence the heavy chains of immunoglobulins or antibodies, they have to be isolated from serum myeloma proteins of patients suffering from multiple myeloma. That was achieved about five years later. They are in general about twice as long as the light chains. For example, one of the first completely sequenced heavy chains, EU (Edelman *et al.*, 1969), has 446 amino acid residues. Another, OU (Putman *et al.*, 1973), has 576 residues. The N-terminal quarter of these sequences are highly varied as the N-terminal half of the light chains. The remaining three-quarter of the heavy chain can roughly be divided into three similar regions. Between the first and second regions, there is usually a short segment rich in Cys residues known as the hinge region.

A schematic representation of an antibody molecule is shown in Fig. 1-2 (see, for example, Kabat *et al.*, 1991). It consists of two heavy and two light chains. As discussed before, the light chain can be divided into a variable, V_L , and a constant, C_L , region, while the heavy chain into V_H , C_H1, hinge, C_H2 and C_H3 regions. In some cases, the hinge region is absent, and there is an additional C_H4 region. Amino acid residues are numbered from the N-terminal ends. Nearly every researcher has a different numbering system. However, alignment will be very important to further analyze these sequences. We are thus going to use the Kabat numbering system (Kabat *et al.*, 1991) which has been generally adopted by many investigators in this field.

Each of us can produce two different types of light chain constant regions, known as kappa or κ and lambda or λ. Their genes are located on two different chromosomes, 2p11 (Malcolm *et al.*, 1982) and 22q11 (Erikson *et al.*, 1981; Emanuel *et al.*, 1985) respectively. We can also produce several different heavy chain constant regions, known as μ, δ, γ, α, ε, etc. and some minor variations of these. Their genes are all located consecutively on the same chromosome, 14q32 (Croce *et al.*, 1979; McBride *et al.*, 1982).

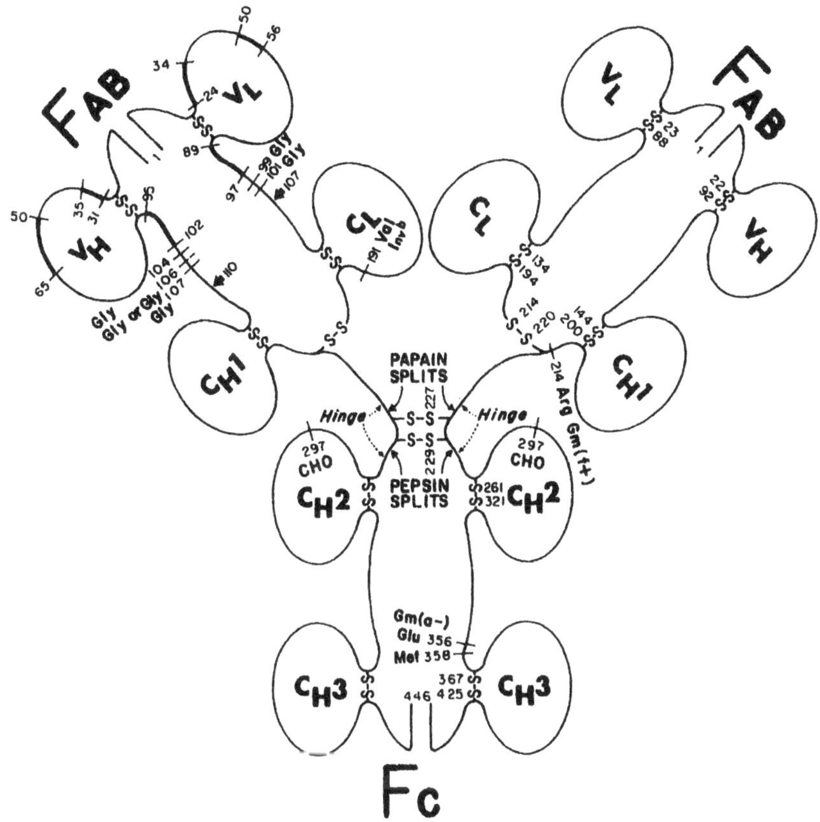

Figure 1-2. Schematic representation of antibody molecule (from Kabat *et al.*, 1991).

The light chains are covalently linked to the heavy chains by disulfide or S-S bonds. Several S-S bonds in the hinge regions of two heavy chains link them together. Since the hinge regions are relative exposed to the solvent, enzymes, such as papain or pepsin, can cut the peptide backbones (Porter, 1959). After such digestion, the antibody molecule is broken up into two Fab fragments and one Fc fragment. The Fc portion is the relatively constant end of the antibody molecule capable of binding special Fc receptors on macrophages and other cells able to engulf foreign antigens. The Fab portions of antibodies will bind specific antigens. The specificity resides in the amino acid sequences of the V_L and V_H regions collectively known as the F_V fragment of the Fab portion of the antibody molecule.

Therefore, to understand the adaptive aspect of our immune system, we need to determine a large number of amino acid sequences of the variable regions of the light and heavy chains of antibodies. The relative simplicity of isolating Bence Jones proteins from urine samples of multiple myeloma patients provided a huge experimental resource of such sequence studies. Bence Jones proteins had been used as a diagnostic method for multiple myeloma. In the late 1960's and early 1970's these proteins have provided vast amount of experimental data to our understanding of the intricacies of our antibody system.

Within five years, around eighty partial or complete amino acid sequences of Bence Jones proteins from human patients were determined. A couple mouse sequences were also available. The mouse model for human multiple myeloma was developed by Potter (1968) by injecting mineral oil into the peritoneal cavity of some inbred strains. Their V_L regions consisted of about 107 amino acid residues as noted by Hilschmann and Craig (1965). It was immediately clear that we could not have enough genetic material to code for millions or billions of different variable region sequences separately. Furthermore, one of the central problems was how different antibodies could have the ability to bind specific antigens.

In order to study this problem, these sequences were aligned, by positioning the two invariant Cys residues forming an internal S-S bond at fixed locations numbered 23 and 88 from the N-terminal. An invariant Trp residue was fixed at position 35, and two invariant Gly residues at positions 99 and 101. Some sequences were longer, and gaps were introduced between positions 27 and 28, and between 95 and 96 (Figure 1-1). This scheme is the Kabat numbering system for the variable regions of antibody light chains (Kabat *et al.*, 1991).

My friend, Dr. Kabat, and I were staring at such alignments on long strips of yellow papers placed on a large conference table, and realized that we had to introduce some quantitative measure. Biochemists have long realized that for most of the proteins, there are only twenty different amino acid residues linked linearly. We reasoned that if the lengths of the variable regions of light chains were so similar, their segments responsible for the binding of antigens would consist of different amino acid residues from sequence to sequence. This situation is in complete contrast to proteins with one specific function. For example, cytochrome c's from different species have different sequences. However, since they all serve the same function of electron transport, their active sites should consist of similar or identical amino acid residues.

Our analysis had to be simple. Otherwise, immunologists or molecular biologists would not appreciate our findings. After many attempts, we settled on the following ratio as a function of position, and named it variability (Wu and Kabat, 1970):

$$V(P) = N / D ,$$

where V is the variability, a function of position P. For a set of aligned amino acid sequences, the numerator N is the number of different amino acid residues found at that position. The denominator D is the frequency of the most common amino acid residue at that position. For example, if at position 23, Cys is found in all sequences, we have $N = 1$, and $D = 1$. Thus,

$$V(23) = 1 / 1 = 1 .$$

Therefore, variability is equal to one for an invariant position. On the other hand, the theoretical maximum of V is for a position P where all twenty amino acid residues are found in different sequences, and they occur at the same frequency, i.e. $N = 20$, and $D = 1/20$. Then, for that position P, we have:

$$V(P) = 20 / (1/20) = 400 .$$

For actual collections of aligned experimentally determined sequences, this maximum value is never reached. However, the large range of 1 to 400 gives us sufficient flexibility to analyze the extent of variation of amino acid residue substitutions at various positions.

One may ask what the big deal is for this "trivial" analysis. We first applied this calculation to a collection of sequences of cytochrome c's from different species provided to us by Dr. Margoliash. Most of them consist of 104 amino acid residues, similar to the length of the variable region of light chains of immunoglobulins. The plot is shown in Fig. 1-3.

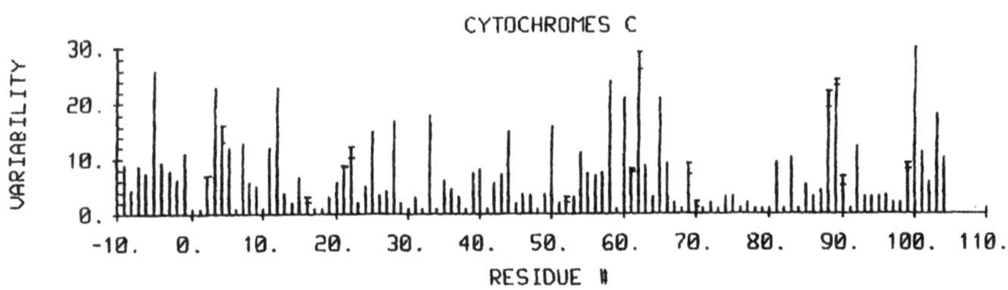

Figure 1-3. Variability plot for cytochrome c's (Kabat *et al.*, 1991).

Indeed, there are many positions with variability equal to one. Furthermore, the maximum variability is around 30, much smaller than 400.

The interesting part was that for the 107 positions of the light chain variable region, a plot of variability V against position P showed three distinct peaks, at positions 24 to 34, 50 to 56, and 89 to 97. That was totally unexpected. But it was subsequently verified repeatedly, not only for light chains but also for heavy chains where the peaks are located at positions 31 to 35, 50 to 65, and 95 to 102. Representative plots are shown in Figs. 1-4 and 1-5.

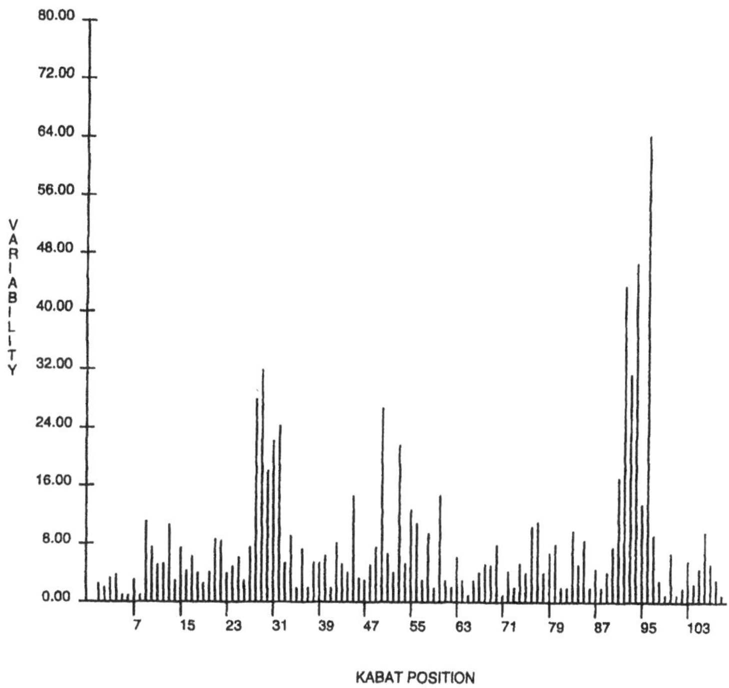

Figure 1-4. Variability plot for some antibody (human rheumatoid factors) κ light chain variable region sequences.

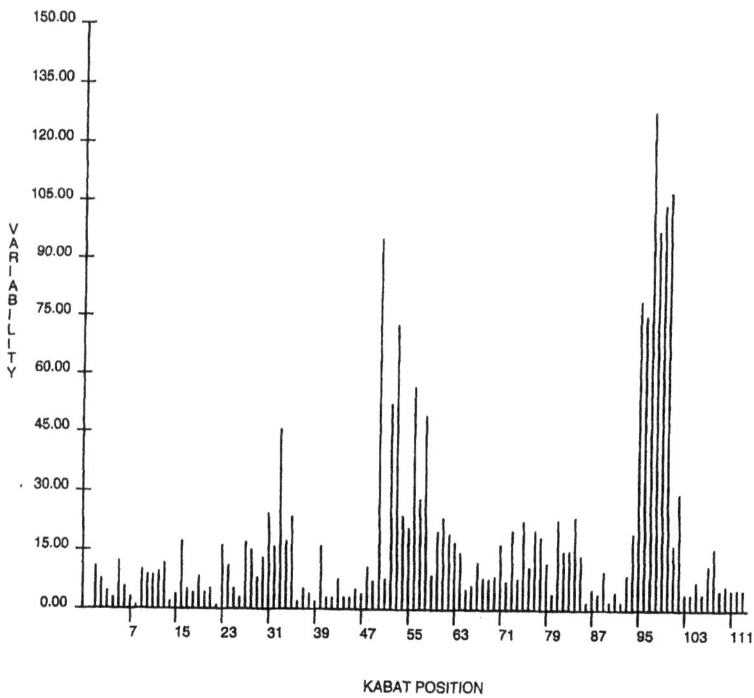

Figure 1-5. Variability plot for some antibody (human rheumatoid factors) heavy chain variable region sequences.

Immunologists, as well as biochemists, molecular biologists, and other biologists, simply could not believe this simple result. We proposed that the variable regions of antibody light and heavy chains would fold together with these six short segments on one side of this globular structure forming the antibody combining site. These segments were thus designed as complementarity determining regions (CDR's). Similar to what the enzymologists had proposed for the binding of enzyme molecules with their substrates in a lock-key configuration, we proposed the same for antibodies binding antigens. Furthermore, various combinations of these six "fingers" could easily generate the required number of millions or billions of different antibodies. Using the Kabat numbering system, we specify the positions for the six CDR's as listed in Table 1-1. Other portions of the variable region are designated as framework regions (FR's). Slight variations are observed in some species.

Table 1-1. Positions for FR's and CDR's of light and heavy chains of antibodies (modified from Kabat *et al.*, 1991).

Segment	Light Chain	Heavy Chain
FR1	1-23 (with occasional 0, and deletion of 10 in V_λ)	1-30 (with occasional 0)
CDR1	24-34 (with possible insertions between 27 and 28)	31-35 (with possible insertions between 35 and 36)
FR2	35-49	36-49
CDR2	50-56	50-65 (with possible insertions between 52 and 53)
FR3	57-88	66-94 (with possible insertions between 82 and 83)
CDR3	89-97 (with possible insertions between 95 and 96)	95-102 (with possible insertions between 100 and 101)
FR4	98-107 (with possible extra residue after 106)	103-113

Depending on the total number of sequences used in the calculation, CDR's have higher values of variability, about 50 to 150, and FR's lower values, less than 30 as in the case of cytochrome c's. In addition, amino acid residues just in front and just after CDR's are usually invariant or nearly invariant. Even though nucleotide sequences of antibody genes would not be determined until eight years later, some basic mechanisms of antibody production became imaginable. During the subsequent 30 years, the rapid development of genetic and protein engineering technology opened up the possibility of artificially synthesizing designer antibodies for therapeutic uses.

VERIFICATION OF PREDICTIONS

Some of the immunoglobulins from mouse models of multiple myeloma were soon isolated, purified and crystallized. However, what antigens these immunoglobulins would bind were not known. Their three dimensional structures, as determined by X-ray diffraction studies, showed that the six CDR's were indeed physically located together on one surface of the molecule. The third CDR's of both light and heavy chains were in the middle of the site, while the second CDR's on the periphery. They could indeed bind foreign antigen molecules.

The actual verification of our prediction would, however, wait for the development of a new technology, namely the production of monoclonal antibodies. Kohler and Milstein (1975) developed a method of producing antibodies of a defined specificity in a continuous culture. Thus, when a mouse is injected with a specific antigen, e.g. lysozyme, its spleen cells can be fused with myeloma cells to generate such continuous cultures known as hybridoma cell lines. The antibody produced by each cell line is uniform, like the myeloma proteins, except that its antigen is known. Usually, various anti-lysozyme antibodies with different binding affinities can be isolated. In most cases, complete antibodies are difficult to crystallize due to the flexible nature of their hinge regions. Thus, their Fab fragments are manufactured by papain digestion (Porter, 1959), and are co-crystallized with the antigen molecule. If the binding affinity of the Fab fragment for lysozyme is sufficient high, they would form stable crystals. Eventually, 15 years later, with the development of the this technology, an anti-lysozyme antibody Fab fragment was co-crystallized with lysozyme (Fig. 1-6), so that their contacting amino acid residues could be analyzed in detail (Amit *et al.*, 1986).

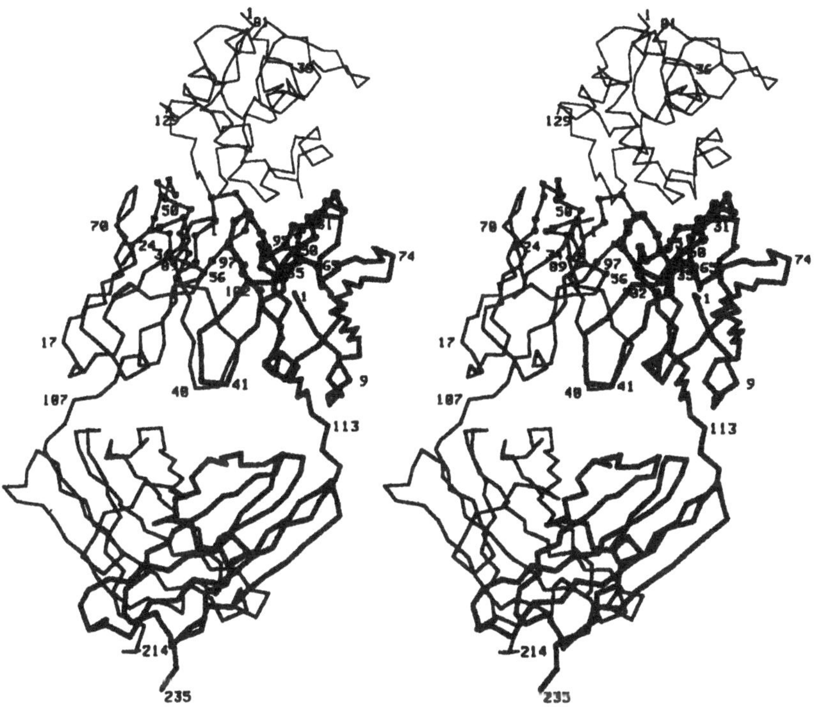

Figure 1-6. Stereo-pair of an anti-lysozyme antibody Fab fragment binding lysozyme (from Kabat *et al.*, 1991).

Indeed, amino acid residues from all six CDR's participate in the binding of the antigen molecule. Several immediately adjacent residues are also involved (Table 1-2).

Table 1-2. Amino acid residues of an anti-lysozyme antibody in direct contact with lysozyme (from Kabat *et al.*, 1991).

Light Chain	CDR1	His 30
		Tyr 32
	FR2	Tyr 49
	CDR2	Tyr 50
	CDR3	Phe 91
		Trp 92
		Ser 93
Heavy Chain	FR1	Thr 30
	CDR1	Gly 31
		Tyr 32
	CDR2	Trp 52
		Gly 53
		Asp 54
	CDR3	Arg 96
		Asp 97
		Tyr 98
		Arg 99

The idea of how the combination of various CDR's can give rise to antibodies of different specificities, and the identification of numerous antibody genes became one the most fascinating area of molecular biology research during the past 20 years. A brief summary of the generation of CDR3's of the antibody light and heavy chains will be discussed below.

SWITCH PEPTIDE AND J-MINIGENE

In 1967, Milstein was trying to estimate the number of genes for human kappa light chains. He noticed that in the variable region, subgroups could be defined by using the first 23 amino acid residues from the N-terminal. However, there were several different sequences in the region from position 104 to 107 for each subgroup, and these short segments were shared by various subgroups. He designated them as switch peptides. We (Kabat *et al.*, 1978) made a detailed analysis of the amino acid sequences between the CDR's, i.e. the FR's, and noted that some of these FR amino acid sequences were identical among human, mouse and rabbit. For different antibodies with different CDR's, their FR's can be identical, suggesting that the variable regions of antibody light and heavy chains can be assembled from short segments.

In 1978, Bernard *et al.* were analyzing the nucleotide sequences of mouse lambda light chain genes in adult and fetus. They discovered that the light chain variable region gene was assembled from the joining of a V-gene and a J-minigene, which were physically separated by about 100 kb in the fetal genome. This process is now known as DNA re-arrangement. In fact, the J-minigenes code for the amino acid sequences of switch peptides. Special recombination signal sequences (RSS), consisting of a 7-mer and a 9-mer separated by a spacer of around 12 or 23 base-pairs, are involved in this re-arrangement (Fig. 1-7). A 23-bp spacer at the 3'-end of the mouse lambda light chain V-gene can be joined to a 12-bp spacer at the 5'-end of the J-mingene. Many different enzymes are required (Slackman *et al.*, 1996). Portions of the nucleotide sequences for the V-gene and J-minigenes of the mouse lambda light chain, and their RSS are shown in Fig. 1-7. For kappa light chains, the spacers for their RSS are reversed, i.e. a 12-bp spacer at the 3'-end of the V-gene and a 23-bp spacer at the 5'-end of the J-minigene.

```
Fetal   tat ttc tgt gct cta tgg tac agc aac cat ttccacaatgacatgtgtagatggggaagtagatcaagaaca
                                                --------                   -----------
                                                7-mer       23-bp          9-mer

Fetal          ggtttttgcatgagtctatatcacagtgc tgg gtg ttc ggt gga gga acc aaa ctg act gtc cta g
               ----------       ---------
               9-mer     12-bp  7-mer

Adult   tat ttc tgt gct cta tgg tac agc aac cat tgg gtg ttc ggt gga gga acc aaa ctg act gtc cta g

                    V-gene                             J-minigene
```

Figure 1-7. Portions of nucleotide sequences of the fetal and adult mouse lambda light chain variable region V-gene and J-minigene (modified from Bernard *et al.*, 1978).

The junction occurs within the coding region of the third CDR3 of the light chain variable region (Fig. 1-8). The J-mingene of the mouse lambda light chain was located 1.2 kb upstream from the 5'-end of the exon of the constant region. DNA re-arrangement brought the exon of the variable region close to that of the constant region.

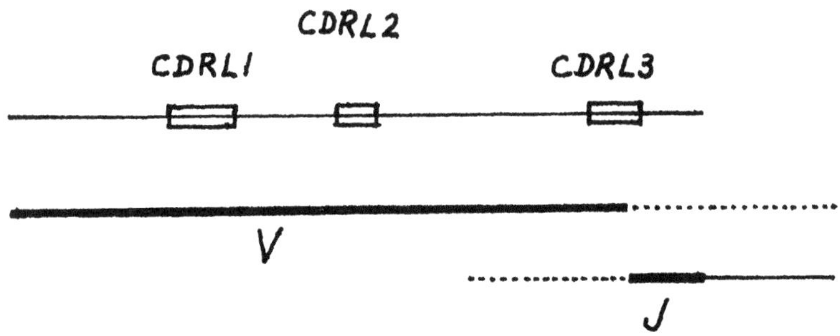

Figure 1-8. Relative locations of CDR's in the light chain variable region to the V-gene and J-minigene.

As shown in Fig. 1-7, the joining of the V-gene and the J-minigene has some flexibility. It does not occur at a precise location. Thus, different codons can result to give rise to different amino acid residues, known as junctional diversity. The presence of many different amino acid residues at this junction automatically includes that position under the third CDR of the light chains (Fig. 1-8). However, the reading frames of both V-gene and J-minigene have to be maintained, in order to produce a useful protein. Otherwise, the re-arrangement V-J segment is designated as a pseudo-gene. The inclusion of other additional nucleotides at this junction is relatively rare, except possibly for rabbit kappa light chains (Sehgal *et al.*, 1999).

Soon afterwards, other researchers discovered that for heavy chain variable region, the situation is more complicated (Early *et al.*, 1980; Sakano *et al.*, 1980). The variable region gene of heavy chain is assembled from the joining of a V-gene, a D-minigene and a J-minigene. Furthermore, the addition of possible intervening sequences at their junctions is quite frequent. They are known as N-segments, if G,C-rich (Alt and Baltimore, 1982), or P-segment, if A,T-rich (Lafaille *et al.*, 1989). All of these occur in the segment coding for the third complementarity determining region of the heavy chain, CDRH3 (Fig. 1-9). The J-minigenes of heavy chain variable region are located about 6 kb upstream from the 5'-end of the exon of C_H1 of the μ chain, IgM.

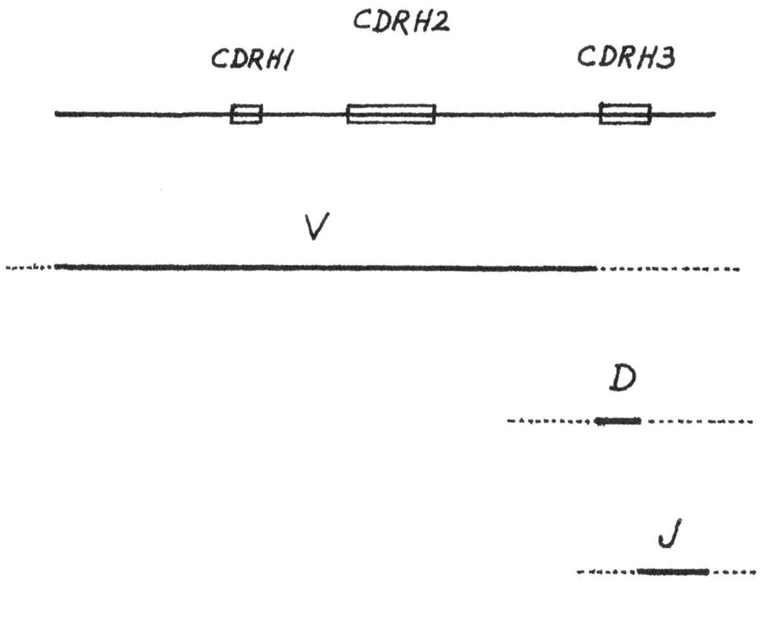

Figure 1-9. Relative locations of CDR's in the heavy chain variable region to the V-gene, D-minigene and J-minigene.

UNIQUENESS OF CDRH3

Assuming that the six CDR's can be randomly assorted to generate antibodies capable of binding various antigens, we would expect that each specific CDR amino acid sequence would be associated with several different specificities. This is indeed the case for CDRL1, CDRL2, CDRL3, CDRH1 and CDRH2. However, a given CDRH3 sequence with five amino acid residues or more is always associated with a unique specificity (Table 1-3). Several representative amino acid sequences are illustrated here (Wu, 1994).

Table 1-3. Association of certain amino acid sequences of CDR's with different antibody specificities (modified from Wu, 1994).

CDR	Amino acid sequence	Number of different specificities
CDRL1	RSSQSLVHSNGNTYLS	16
CDRL2	KVSNRFS	29
CDRL3	SQSTHVPWT	4
CDRH1	DYYMN	10
CDRH2	DINPNNGGTSYNQKFKG	8
CDRH3	YYYGSSLV	1

Since CDRH3 is located in the middle of antibody combining sites, it seems that this segment of an antibody molecule effectively defines its fine specificity. In human, CDRH3 can vary from just a few amino acid residues in length to 31 residues. In cows, camels, etc., it can even be longer. In mouse, however, the longest CDRH3 is 19 amino acid residues for sequence so far determined.

Thus, if we want to design an antibody with a special property, we should start from selecting a CDRH3. Most of them form loop structures, with about seven or eight amino acid residues giving a more or less flat surface to the antibody combining site. Shorter ones provide a recessed surface, while longer ones a protrusion. Length distributions for human and mouse CDRH3's are illustrated in Fig. 1-10. In our database available from the website <http://immuno.bme.nwu.edu> (Johnson and Wu, 2000), we have collected a total of over 2,500 complete and distinct sequences of human and mouse CDRH3's each. Interestingly enough, there is only one complete amino acid sequence of CDRH3 shared by human and mouse. It is:

Gly Leu Ser Gly Phe Asp Tyr

Their nucleotide sequences are different. Are CDRH3's species-specific? We do not know.

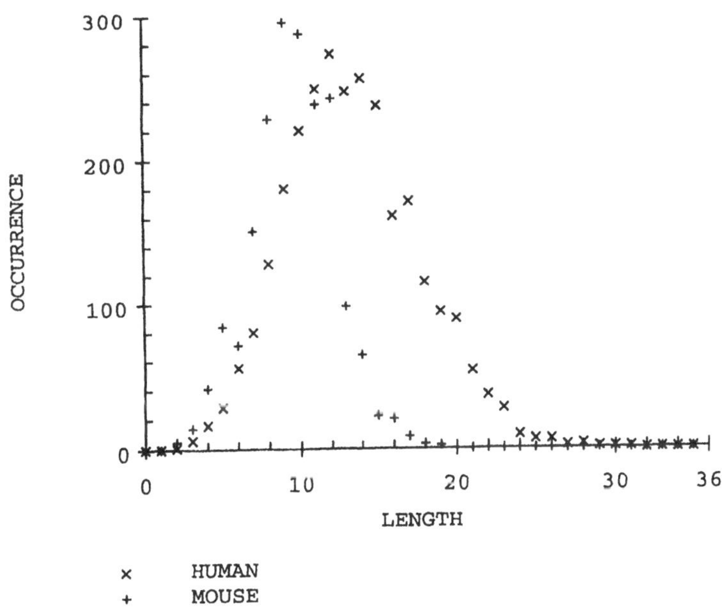

Figure 1-10. Length distributions of human and mouse CDRH3's.

This uniqueness of CDRH3 sequences, especially at the nucleotide level, has opened up a powerful method of detecting residual diseases of B cell lymphomas. The heavy chain variable region nucleotide sequences from such patients are determined before the initiation of treatment. The CDRH3 sequences are then synthesized and used as probes to detect the existence of a very small number of malignant cells. Treatment will be continued until no such cells can be found by this sensitive method. Cross-reaction of the probe with other heavy chain sequences is usually minimal.

CODING OF CDRH3

Among the above mentioned six CDR's, the coding segment for CDRH3 is derived completely differently from the other five. Usually, a D-minigene is first joined to a J-minigene by the process of re-arrangement involving a number of enzymes (Slackman *et al.*, 1996). The joining is not precise, and intervening nucleotides can be added. Probably two different mechanisms are involved to incorporate these additional N/P segments. Subsequently, this D-J segment is then joined to a V-gene, using the same collection of enzymatic reactions. CDRH3 starts two amino acid residues away from the second invariant Cys, and ends one residue from the Gly ( ) Gly pair. According to the Kabat numbering system, CDRH3 is from position 95 to 102, with possible insertions between positions 100 and 101. A schematic drawing is illustrated in Fig. 1-11.

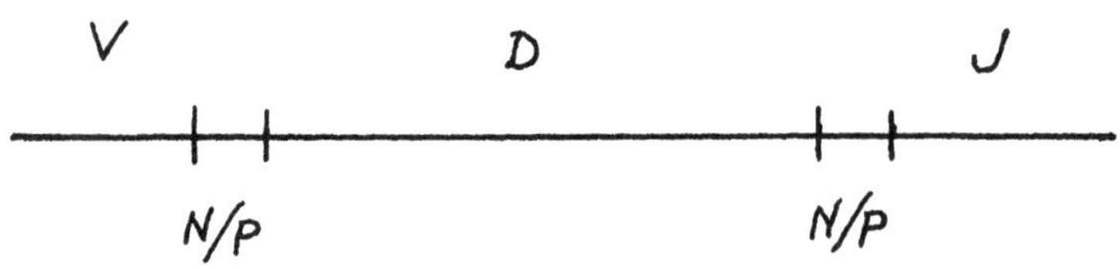

Figure 1-11. CDRH3 consists of the 5'-end of V-gene, N/P segment, D-minigene, another N/P segment, and the 5'-end of J-minigene.

In order to maintain the reading frames in the V-gene and the J-minigene, so that the resulting protein can be functional, the total number of nucleotides of the segment coding for CDRH3 must be divisible by three. However, the number of nucleotides of the N/P segments on both ends of the D-minigene can vary. As a result, all three reading frames of any D-minigene can be used, as long as there is no termination codon. Furthermore, since the recombination signal sequences on both sides of the D-minigene to join D to J and V to D are similar, D-minigenes can be inverted. Thus, the complementary strands of D-minigenes can also be used to code for CDRH3. In human and mouse, however, the complementary strands are found less frequently in the heavy chains of antibodies. In short, one nucleotide sequence of a D-minigene has the capacity of coding for six different amino acid sequences (Fig. 1-12).

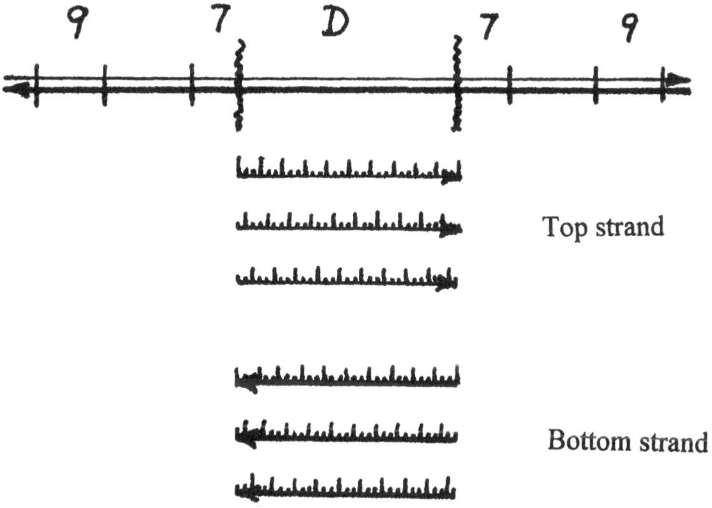

Figure 1-12. One nucleotide sequence of a D-minigene can code for six different amino acid sequences

Some of the longer D-minigenes may have, in their coding regions, a short segment similar to the 7-mer of the recombination signal sequence. In certain cases, therefore, such D-minigenes can then join onto another D-minigene, giving rise to longer nucleotide sequence for coding (Fig. 1-13). Indeed, some of the longer CDRH3 sequences consisted of D-D joining (Kurosawa and Tonegawa, 1982), and even D-D-D joining. The combination of all of the above mentioned mechanisms of generating coding regions for CDRH3 can in fact result in thousands, millions or even billions of different sequences of varying amino acid residues and lengths. It has thus been suggested that CDRH3 alone is sufficient to define the fine specificity of an antibody. The other five CDR's would improve the overall binding affinity of the antibody.

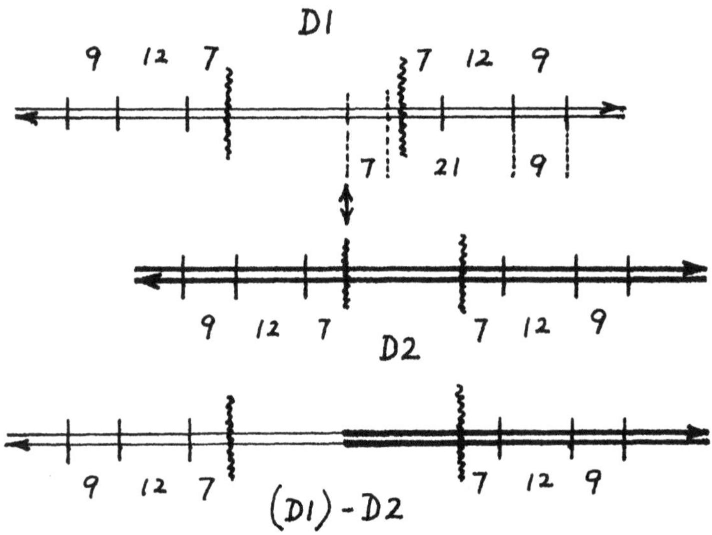

Figure 1-13. The mechanism of D-D joining.

VACCINATION

Our ability to produce millions or billions of different antibodies was used to develop the most powerful tool to prevent infectious diseases, namely vaccination. Jenner was credited for the discovery of vaccination (Osler, 1918), although it was actually employed by a Dorsetshire farmer to protect his family from small pox infection with cow pox (vaccinia). The idea is that some protein or other substances on less virulent or attenuated bacterial or viral agents can stimulate our immune system to produce specific antibodies against them. Subsequent infections of similar or pathogenic bacteria or viruses can result in the binding of our antibodies specifically to these bacteria or viruses to promote their being engulfed by macrophages, leukocytes, etc. For each antigen, the production of specific antibodies is on random basis. Some of the antibodies would have high binding affinities, while others not so high. Furthermore, the duration of production of these antibodies varies. Thus, the standard practice is that periodically, booster vaccination will be required.

On the other hand, some infectious agents, e.g. the flu virus, the human immunodeficiency virus (HIV), etc., can mutate their surface protein molecules thus avoiding their binding to our antibodies. It is for this reason that flu vaccination is required yearly for different strains of viruses. An effective vaccine for HIV is still under development. Recently, DNA segments coding for some of the surface proteins of HIV have been used for vaccination. Hopefully, these segments can be incorporated into our chromosomes and direct the synthesis of specific viral proteins on continuous basis. In particular, these segments have been used to replace the CDRH3 of human antibody genes, so that such pieces of viral proteins are indeed presented on the surface of a protein in a loop configuration.

Another method of preventing bacterial or viral infection is passive immunity. Antibodies from a donor, usually in the form of serum gamma globulin, can be injected into recipients for temporary protection. This is sometimes used for travelers to some countries with a number of prevalent infectious diseases. Serum gamma globulin probably contains millions of different antibodies, a few of which may be particularly useful to combat these infectious diseases.

The above mentioned techniques have lately been extended to the treatment of cancer. The idea is that to a certain extent cancer cells are foreign. Presumably, some of their cell surface molecules are different from those on normal cells or are extra abundant. Thus, specific antibodies may be raised against such molecules. Eventually, when the basic interactions between antibodies and antigens are understood at the molecular level, we should be able to design specific antibodies to a given antigen. However, at present, our analytical knowledge is still lacking. We thus have to solve this problem with a different approach as discussed below.

CHIMERIC ANTIBODIES

Before the days of genetic and protein engineering, some mouse antibodies had been tried in human patients to combat infection. It was noticed that those mouse antibodies with the Cμ heavy chains, or IgM, had longer half-life in human patients than those with other types of heavy chain constant regions. The reason might be due to the relative similarity between the amino acid sequences between human and mouse IgM's. Although longer, the half-life of mouse IgM's was still too short for effective therapeutic use. The basic trouble was that human patients would consider mouse antibodies as foreign antigens, and develop anti-antibodies to get rid of them.

The antigen binding part of antibodies is located in the light and heavy chain variable regions, the V_L and V_H. Therefore, the immediate idea to make mouse antibodies more tolerable in human patients is to graft the mouse V_L and V_H to human light and heavy chain constant regions respectively (Fig. 1-14). Thus, the only antigenic portions of such chimeric antibodies are the

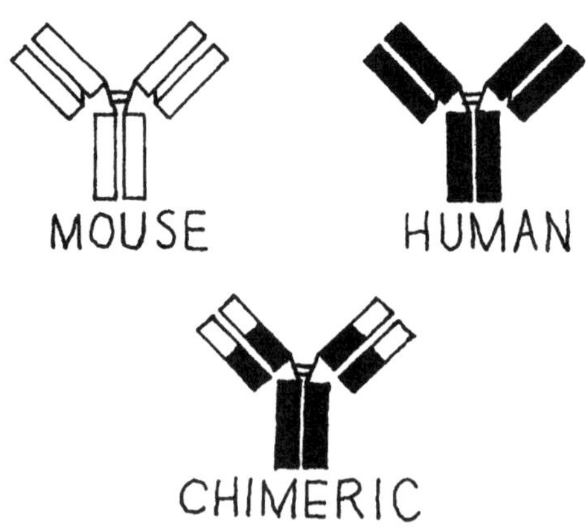

Figure 1-14. Schematic representation of a chimeric antibody.

mouse light and heavy chain variable regions. Their half-life in human patients can be around two weeks, and may exert their biological effectiveness during that period. One of the first approved therapeutic antibodies of this type is an anti-CD20 antibody, designed to treat low grade B cell lymphoma. CD20 is a surface molecule present in most B cells. Various pharmaceutical companies are developing other chimeric antibodies of hybrid mouse and human origin to treat cancers and viral infections.

HUMANIZING MOUSE ANTIBODIES

Chimeric antibodies make use of V_L and V_H of mouse antibodies containing the CDR's selected for certain specificities. In addition they also contain the framework regions (FR's) of the original mouse antibodies. Some of these FR's can be antigenic, thus reducing their half-lives in human patients, even though several FR amino acid sequences are conserved among human, mouse and rabbit. At the gene level, it is possible to cut out the segments coding for CDR's from mouse antibody genes. These segments can then be grafted onto gene segments coding for human FR's. The resulting antibody is known as humanized mouse antibody. The major question in manufacturing such antibodies is which human FR's should be used.

Ideally, in order to maintain the avidity of a humanized mouse antibody, the mouse FR's should be matched against all known human FR's. The most similar human FR's should then be used. This can preserve the overall three-dimensional folding of the resulting antibody molecule. To achieve this, a special program for sequence matching has been included in our website, <http://immuno.bme.nwu.edu>. An example is illustrated in Fig. 1-15. On opening our website, click on SeqhuntII. A new window will open up. Pick Match. In the third window, pick amino acid and choose HUMAN IG HEAVY chain variable region, for example. Then, type in the required amino acid sequence in single letter abbreviation. The example shows the match of a mouse heavy chain FRH1 sequence of V11, i.e.

EVKLVESGGGLVQPGGSLRLSCATSGFTFT

There are four best matches for human heavy chain FRH1 with two amino acid residue differences. One of them is mAB55 as shown in Fig. 1-15.

```
HUMAN IG HEAVY CHAIN VARIABLE REGION

NAME   mAb55

REF    IKEMATSU,H.,HARINDRANATH,N.,
       UEKI,Y.,NOTKINS,A.L. & CASALI,
       P. J IMMUNOL 150: 1323-1337
       (1993)

SEQ         1      GLU   GLU
            2      VAL   VAL
            3      GLN   lys
            4      LEU   LEU
            5      VAL   VAL
            6      GLU   GLU
            7      SER   SER
            8      GLY   GLY
            9      GLY   GLY
           10      GLY   GLY
           11      LEU   LEU
           12      VAL   VAL
           13      GLN   GLN
           14      PRO   PRO
           15      GLY   GLY
           16      GLY   GLY
           17      SER   SER
           18      LEU   LEU
           19      ARG   ARG
           20      LEU   LEU
           21      SER   SER
           22      CYS   CYS
           23      ALA   ALA
           24      THR   THR
           25      SER   SER
           26      GLY   GLY
           27      PHE   PHE
           28      THR   THR
           29      PHE   PHE
           30      SER   thr
```

Figure 1-15. Matching a mouse FRH1 with human sequences.

HUMAN ANTIBODIES IN MOUSE

As we have discussed before, among more than 2,500 human and mouse CDRH3 amino acid sequences each, only one sequence is shared by both. The question is then whether mouse CDR's are also antigenic in human patients. Even though the half-lives of humanized mouse antibodies is approaching that of human antibodies in human patients, they are still not the same. Thus, recently, transgenic mice carrying human antibody genes have been constructed. These animals may thus be used to selectively produce human antibodies with required specificities. However, whether such transgenic mice use preferably mouse or human genes for the production of their antibodies requires careful analysis. In other words, it will be essential to identify the origin of their CDR's. Our finding of nearly distinct sets of human and mouse CDR's can provide some information by sequence comparison as illustrated above.

PHAGE DISPLAY LIBRARY

Some filamentous phages can carry extra pieces of DNA. Human antibody variable regions genes, i.e. V-genes and J-minigenes of the light chains, and V-genes, D-minigenes and J-minigenes of the heavy chains, can be incorporated. By various combinations, different phages will display different V_L and V_H on their surfaces. These V_L and V_H will associate to form globular structures known as F_V with the same specificities of Fab. Since large numbers of phages can be processed and screened, those displaying very high binding affinity F_V can be selected. Separate libraries can be constructed for human genes, for mouse genes, etc.

After the V_L and V_H genes have been selected for specified binding properties, they can then be joined onto the constant region genes of the light and heavy chains respectively for the production of biologically active complete antibody molecules.

In short, there are various biological tools capable of constructing antibodies with desired properties. However, the binding specificity between antibody and antigen molecules is not understood, and left to a random selective process. In order to tackle this basic problem, we have to analyze how CDR's are folded, how the CDRH3 defines the fine specificity of an antibody, how the combination of all six CDR's can improve the binding affinity, etc.

FOLDING OF CDR's AND PREDICTION OF ANTIBODY COMBINING SITES

For a given amino acid sequence of a certain CDR of an antibody, predictive methods will be explained in detail in Chapter 6 of how it may fold in three dimension. Since the framework regions of most antibodies assume a similar structure, they can form a foundation for the predicted CDR's to join onto. Starting from the middle of the combining site, i.e. CDRH3 and CDRL3, we can gradually build the entire structure of the six CDR loops. The site generates a very compact and well-defined surface where the antigen molecule can interact.

ANTIBODY DESIGN

Our understanding of the properties of protein surface structures is still very limited. There may be patches of charged residues, three-dimensional contours of various sizes and shapes, hydrophobic or hydrophilic areas, etc. It will be important for us to describe them mathematically in order to get some insight to the basic mechanisms of antibody-antigen interaction. While we have localized the region of antibody molecules capable of binding antigens, we still have no detailed knowledge of the basic forces involved in such binding. To manufacture designer antibodies to bind predefined antigens will be an extremely challenging area of study for the joint effort of biologists and mathematicians.

APPLICATION TO OTHER PROBLEMS IN MOLECULAR BIOLOGY

Variability plot can also be applied to any collection of homologous proteins. Similar to antibodies, the T cell receptors (TCR's) for antigen can also bind antigens in a somewhat different context. The TCR α and β chain heterodimers bind processed peptides from foreign antigens sitting in the grooves of major histocompatibility complex (MHC) class I or II molecules. On the other hand, TCR γ/δ chains can be MHC-restricted like α/β chains, or can be MHC-unrestricted. Variability plots for human TCR γ and δ chain variable regions are shown in Figs. 1-16 and 1-17 (Johnson and Wu, 2000). Their CDR3's are clearly indicated. However, their CDR1's and CDR2's are uncertain, probably either due to the relatively few known amino acid sequences or due to the less importance of these two CDR's.

The length distributions of the CDR3's of TCR α, β, γ and δ are also very informative. Those for TCR α and β are both very narrow, as well as for their combined length variations (Johnson and Wu, 1999). This can be the consequence of their binding the complexes of processed peptides inside the grooves of MHC class I or II molecules, which are relatively uniform in shape and size. On the other hand, those for γ and δ are broader, especially that for δ showing possibly a bimodal distribution (Fig. 1-18). TCR δ chains with longer CDR3's, in combination with TCR γ chains, may thus bind foreign antigens directly (Melenhorst *et al.*, 1999).

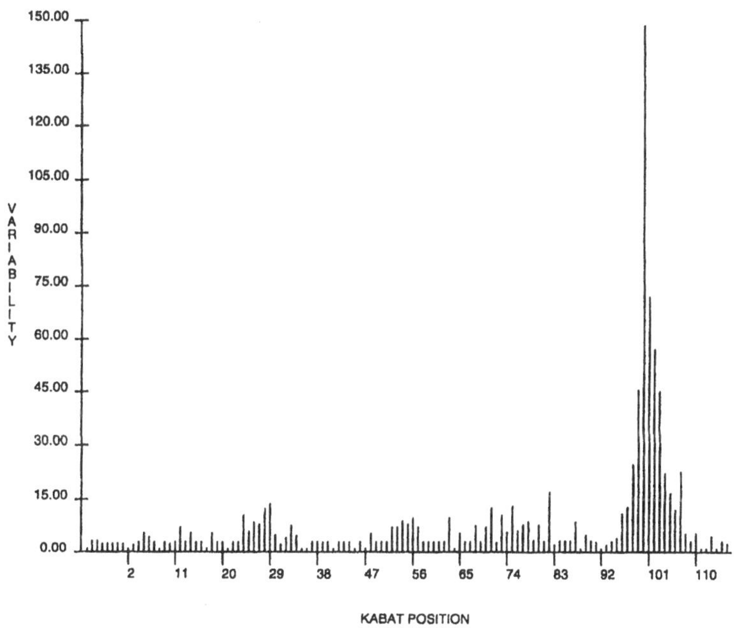

Figure 1-16. Variability plot for human TCR γ chain variable region.

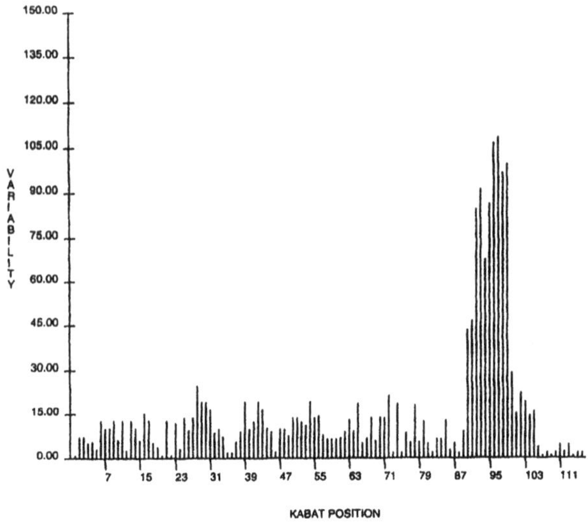

Figure 1-17. Variability plot for human TCR δ chain variable region.

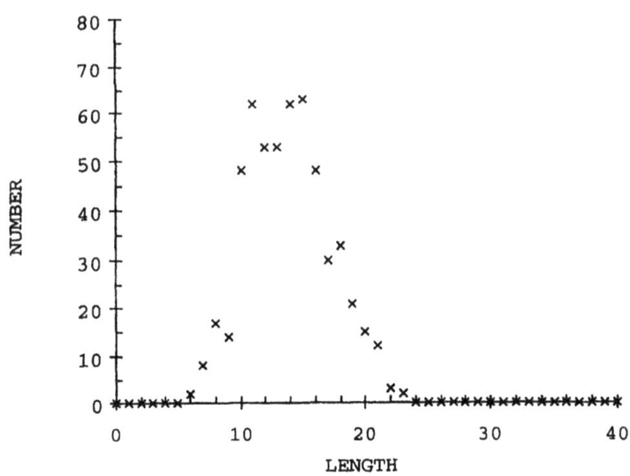

Figure 1-18. Length distribution of human TCR δ chain CDR3's.

Another interesting application of variability plot is the analysis of amino acid sequences of one of the surface glyco-proteins of HIV, gp120. The gene coding for this protein mutates rapidly so that new mutant HIV can avoid the antibodies produced by patients to bind the original virus. Thousands of amino acid sequences of gp120 have been determined and aligned. A variability plot for these sequences clearly identified several segments of relatively higher frequencies of amino acid substitutions. One of these segments, known as the V3-loop, appears distinct from the others. It is this segment that protein and DNA vaccinations are under development as discussed before.

In short, variability plot is one of the simple yet powerful methods of analyzing aligned amino acid sequences of homologous proteins. It can provide essential information about their structure and function relations at a molecular level.

REFERENCES

Alt FW and Baltimore D (1982) Joining of immunoglobulin heavy chain gene segments: implications from a chromosome with evidence of three D-J_H fusions. *Proc. Nat. Acad. Sci. USA,* **79**, 4118-4122.

Amit AG, Mariuzza RA, Phillips SEV and Poljak RJ (1986) Three dimensional structure of antigen-antibody complex at 2.8 A resolution. *Science.* **233**, 747-753.

Bernard O, Hozumi N and Tonegawa S (1978) Sequences of mouse immunoglobulin light chain genes before and after somatic changes.

Croce CM, Shander M, Martinis J, Cicurel L, D'Ancona GG, Dolby TW and Koprowski H (1979) Chromosomal location of the genes for human immunoglobulin heavy chains. *Proc. Nat. Acad. Sci. USA,* **76**, 3416-3419.

Early P, Huang H, Davis M, Calame K and Hood L (1980) An immunoglobulin heavy chain variable region gene is generated from three segments of DNA: V_H , D_H , and J_H . *Cell,* **19**, 981-992.

Edelman GM (1959) Dissociation of γ–globulin. *J. Amer. Chem. Soc.*, **81**,3155-3156.

Edelman GM, Cunningham BA, Gall WE, Gottlieb PD, Rutishauser U and Waxdal MJ (1969) The covalent structure of an entire γG immunoglobulin molecule. *Proc. Nat. Acad. Sci. USA,* **63**, 78-85.

Emanuel BS, Cannizarro LA, Magrath I, Tsujimoto Y, Nowell PC and Croce CM (1985) Chromosomal orientation of the lambda light chain locus: V_λ is proximal to C_λ in 22q11. *Nucl. Acids Res.*, **13**, 381-387.

Erikson J, Martinis J and Croce CM (1981) Assignment of the genes for human immunoglobulin λ chain to chromosome 22. *Nature*, **294**, 173-175.

Hilschmann N and Craig LC (1965) Amino acid sequence study with Bence Jones proteins. *Proc. Nat. Acad. Sci. USA*, **53**, 1403-1409.

Johnson G and Wu TT (1999) Random length assortment of human and mouse T cell receptor for antigen α and β chain CDR3. *Immunol. Cell Biol.*, **77**, 391-394.

Johnson G and Wu TT (2000) Kabat database and its applications: 30 years after the first variability plot. *Nucl. Acids Res.*, **28**, 214-218.

Kabat EA (1976) *Structural Concepts in Immunology and Immunochemistry, 2nd Edition.* Holt, Rinehart and Winston, New York.

Kabat EA, Wu TT and Bilofsky H (1978) Variable region genes for the immunoglobulin framework are assembled from small segments of DNA—A hypothesis. *Proc. Nat. Acad. Sci. USA*, **75**2429-2433.

Kabat EA, Wu TT, Perry HM, Gottesman KS and Foeller C (1991) *Sequences of Proteins of Immunological Interest, 5th Edition.* U.S. Department of Health and Human Services, Public Health Service, National Institutes of Health, NIH Publication No. 91-3242.

Kohler G and Milstein C (1975) Continuous cultures of fused cells secreting antibody of predefined specificity. *Nature*, **256**, 495-497.

Kurosawa Y and Tonegawa S (1982) Organization, structure, and assembly of immunoglobulin heavy chain diversity DNA segments. *J. Exp. Med.*, **155**, 201-218.

Lafaille JJ, DeCloux A, Bonneville M, Takagaki Y and Tonegawa S. (1989) Junctional sequences of T cell receptor γδ T cell lineage and for a novel intermediate of VDJ loining. *Cell*, **59**, 859-870.

Malcolm S, Barton P, Murphy C, Ferguson-Smith MA, Bentley DL and Rabbitts TH (1982) Localization of human immunoglobulin κ light chain variable region genes to the short arm of chromosome 2 by *in situ* hybridization. *Proc. Nat. Acad. Sci. USA*, **79**, 4957-4961.

McBride OW, Battey J, Hollis GF, Swan DC, Siebenlist V and Leder P. (1982) Localization of human variable and constant region immunoglobulin heavy chain genes on subtelomeric band q32 of chromosome 14. *Nucl. Acids Res.*, **10**, 8155-8170.

Melenhorst JJ, van Luxemburg-Heijs SA, Landegent JE, Willemze R, Fibbe WE and Falkenburg JH (1999) Aplastic anemia in donor cells 14 years after bone-marrow transplant. *Lancet*, **353**, 2037-2038.

Milstein C (1967) Linked groups of residues in immunoglobulin chains. *Nature*, **216**, 330-332.

Osler W (1918) *The Principles and Practice of Medicine.* D. Appleton and Company, New York.

Porter RR (1959) The hydrolysis of rabbit γ–globulin and antibodies with cyrstalline papain. *Biochem. J.*, **73**, 119-127.

Potter M (1968) A resume of the current status of the development of plasma-cell tumors in mice. *Cancer Res.*, **28**, 1891-1896.

Putman FW, Florent G, Paul C, Shinoda T and Shimizu A (1973) Complete amino acid sequence of the mu heavy chain of a human IgM immunoglobulin. *Science*, **182**, 287-291.

Sakano H, Maki R, Kurosawa Y, Roeder W and Tonegawa S (1980) Two types of somatic recomination are necessary for the generation of complete heavy chain genes. *Nature*, **286**, 676-683.

Sehgal D, Johnson G, Wu TT and Mage RG (1999) Generation of the primary antibody repertoire in rabbit: expression of a diverse set of *Igk-V* genes may compensate for limited combinatorial diversity at the heavy chain locus. *Immunogenet.*, **50**, 31-42.

Slackman BP, Gorman JR and Alt FW (1996) Accessibility control of antigen-receptor variable region assembly: role of cis-acting elements. *Annu. Rev. Immunol.*, **14**, 459-481.

Wu TT (1994) From esoteric theory to therapeutic antibodies. *Appl. Biochem. Biotech.*, **47**, 107-118.

Wu TT and Kabat EA (1970) An analysis of the sequences of the variable regions of Bence Jones proteins and myeloma light chains and their implications to antibody complementarity. *J. Exp. Med.*, **132**, 211-250.

EXERCISE

Calculate the variability for each position of the following set of twenty sequences and plot variability against position as listed on the tip of these sequences.

```
1        10        20        30        40        50
|        |         |         |         |         |
EIVLTQPGTLSTLSPGERATLSCRASQVSSSYLAWYQQKPGQAPRLLIYG
MAELTQSPVTLSVSPGERVALSCKASQNINDNLAWYQQKPGQAPRLLIYG
EIVLTQSPGTLSLSPGERATLSCRASQTIGTSIHWYQQPPGQAPRLLIYY
MAELTQSPATLSVSPGERATLSCRASQSVGTNLAWYQQKPGQAPRLLIFD
DIQMTQSPFTLSASVGERVTITCRASQRINSRLAWGQQKPGKAPKLLIYK
MAELTQSPSSLSASVGDRVTITCRATQSVSNFLNWYQQKPGEAPTLLIYD
DVMMTQSPLSLPVTLGQPASISCRSSQDGNTYLSWFRQRPGRSPRRLIYN
EIVLTQSPGTLSLSPGERATLSCRASQVSSSYLAWYQQKPGQAPRLLIYG
EFVLTQSPGTLSLSPGDRATLSCRASQVDNYYLGWYQQKPGQAPRLLIFG
MAELTQSPGTLSLSPGERATLSCRASQVSNTFLAWYQQKPGQAPRLLIYD
EIVLTQSPATLSLSPGERATLSCRASEYGNSFMHWYQQKPGQAPRLLIYR
MAELTQAPATLSVSPGERATLSCRASQSVGTNLAWYQQKPGQAPRLLIFD
DVVMTQSPSTVSASVGDRVTLTCRASQSISRWLAWYQQKPGQPPKLLIYW
DIQMTQSPSSVSASVGDRVTIACRASQDISDRLAWYQQKPGKVPKVLIYG
DIQMTQSPSSVSASVGDRVTVTCRASQGISSWLAWYQQKPGKAPKLLIHA
DIKMTQSPSSLSASVGDRVTLTCRASQNVYDSLNWYQQKPGKAPNLLIYG
DIQMTQSPSSLSASVGDRVTITCRASQGIRNDLGWYQQKPGKAPKLLIFA
DIQMTQSPSTLSASIGDRVTITCRASQNIDSWLAWFQHKPGKAPKPLIYG
ELQMTQSPSSLSASVGDRATVSCQASQSIYNYLNWYQQKPGKSPKFLTYR
DIQMTQSPSSLSASVGDRVTITCKSSQNQKIYLAWYQQKPGKAPKLLIYW
```

```
51        60        70        80        90        100
|         |         |         |         |         |
ASSRATGIPDRFSGSGSGTDFTLTISRLEPEDFAVYYCQQYGSSPYTFGQ
ASSRATGIPDRFSGSGSGTDFTLTITRLEPEDFAVYYCQQYGGSPYTFGQ
ASESISGIPDRFSGSGSGTDFTLTISRLEPEDFAVYYCQQSSSWPLTFGQ
ASTRDTYIPDTFSGSGSGTDRALTISSLQSEDFGFYYCQQYDNWPPTFGQ
ASTLQSGVPSRFSGSGSGTDFTLTINSLQPDDFATYYCQQYHTYAWTFGQ
ASTSQSGVPSRFSGSGSGMDFSLTISSLQPEDLAMYYCQASINTPALFGQ
VSKRDSGVSDRFSGSGSGTDFTLKISRVKAEDIGTYYCMQDTYWPWIFGQ
ASSKATGIPDRFSGSGSGTDFTLTISRLEPEDFAVYYCQQYGSSPYTFGQ
ASSRATGIPDRFSGSGSGTDFTLTISRLEPEDFAIYYCQHYGSSPWTFGQ
TSSRATGVPHRFTGSGSGTDFTLTISSLQSEDFAVYYCQQYGYSALTFGG
ASNLETGIPARFSGSGSRTDFTLTYSSLEPEDFAVYYCQQSNEDPRTFGG
ASTRDTYIPDTFSGSGSGTDFALTISSLQSEDFGFYYCQQYDNWPPTFGQ
ASTRKSGVPDRFSGSGSGTDFTLTISSLQADDFATYYCQHYDSFSPSFGQ
ASSLQSGVPSRFSGSGSGTDFTLTINSLQPEDFATYYCQQANSFPLTFGG
ASSLQSGVPSRFIGSGSGTDFTLTITSLQAEDFATYYCQQADSLPPTFGQ
ASYLHSGVPSRFSGSGSGTEFTLTITSLHPEDFATYFCQQGFSDARSFGG
ASSLQSGVPSRFSGSGSGTKFTLTISSLQPEDFATYYCLQHNSYPPTFGQ
AFTLQNGVPSRFSGSGSGTEFSLTISSLQPDDFATYFCQQAHSFPPTFGG
ASSLQRGMPSQFSGSGYGRDFTLTVSSLQPDDRATYYCQQGVRLPYTFGQ
ASTRESGVPSRFSGSGSGTDFTFTISSLQPEDIATYYCQQYYRYPRTFGQ
```

```
101    107
  |      |
  GTKLEIK
  GTKLEIK
  GTKVEIK
  GPKLEVK
  GTKVEIK
  GTRIDMR
  GTRVVFN
  GTKLEIK
  GTKVEMK
  GTKVEIK
  GTKVEEK
  GTKLEVK
  GTKVEIK
  GTKVEMK
  GTKVDFK
  GTKVEIK
  GTKVEIK
  GTRLEIK
  GTKVEIK
  GTKVEIK
```

CHAPTER 2

SATURATION OF HEMOGLOBIN WITH OXYGEN AT EQUILIBRIUM

INTRODUCTION

Human beings, as well as many other animals, have developed a very efficient protein, known as hemoglobin, to carry oxygen from the lungs to the tissues. We also have another similar but simpler protein, known as myoglobin, to store oxygen in our muscles. Due to the presence of large amounts of hemoglobin in our red blood cells, it has been relatively easily purified by biochemists. Similarly, myoglobin can also be purified from muscles. Their biochemistry, physiology, genetics, evolutionary history, three-dimensional structure, etc. have been extensively studied during the past century. Indeed, myoglobin and hemoglobin are the two proteins, the three-dimensional structures of which have first been determined by X-ray diffraction studies. Many genetic diseases have been identified with mutations in the hemoglobin gene. The most well known one is sickle cell anemia.

The binding of myoglobin with oxygen at equilibrium is relatively simple. One oxygen molecule can bind one myoglobin molecule. There is a specific site on the myoglobin molecule containing a smaller molecule, known as heme (Fig. 2-1), with an iron atom in its center capable of binding oxygen molecule reversibly. In this case, the iron atom is not oxidized to the ferric state in the presence of oxygen. However, myoglobin is not efficient enough to transport oxygen from our lungs to our tissues, due to its simple equilibrium kinetics. Probably through gene duplication as well as other evolutionary processes, we have acquired a more complex macromolecule consisting of four subunits, each of which resembles a molecule of myoglobin. The equilibrium and time-dependent kinetics of hemoglobin binding oxygen, as well as various molecular models, have been studied

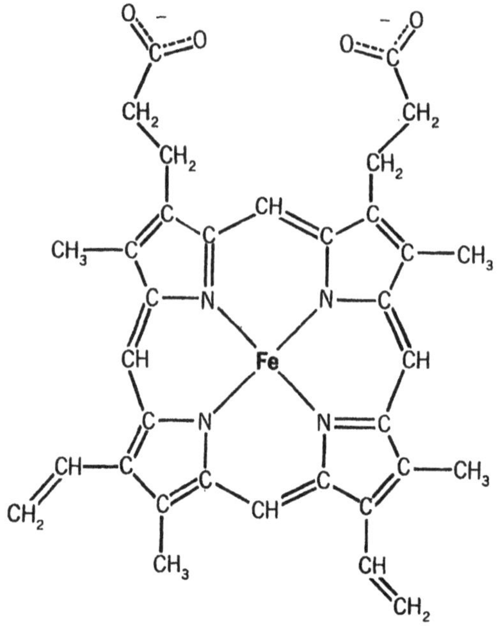

Figure 2-1. The heme.

extensively by numerous famous biochemists and molecular biologists over the years. In this chapter, we are going to review the known biological facts about these two proteins and discuss three models of saturation of hemoglobin with oxygen at equilibrium.

STRUCTURES OF MYOGLOBIN AND HEMOGLOBIN

Perutz has reviewed the structures of myoglobin and hemoglobin in 1962, and, with his collaborators, the cooperative effects of hemoglobin in 1998. His life time contribution to the understanding of these proteins, especially hemoglobin, is monumental. The amino acid sequences of sperm whale myoglobin (Mb) and α and β subunits of human hemoglobin (Hb) are aligned in Table 2-1 (Perutz, 1962), relative to the locations of the helices A through H (Nobbs *et al.*, 1966). The three dimensional structure of myoglobin had been determined before its amino acid sequence was completely characterized. Furthermore, some of the amino acid residues were not visible in the three dimensional structure due to the flexibility of loops connecting helices.

Table 2-1. Amino acid sequences of sperm whale myoglobin and α and β subunits of human hemoglobin (modified from Perutz, 1962).

```
              |-----A-------|     |------B-------||--C--|

Mb               VAGEWSEILKXWAKVQALVAGHGKLTLIRLFKSHPETLEKFDRFKHLK

Hb(α)          VLSPADKTNVKAAWGKVGAHAGEYGAEALERMFLSFPTTKTYFPHF-DLS

Hb(β)         VHLTPEEKSAVTALWGKVN--VDEVGGEALGRLLVVYPWTQRFFESFGDLS

              |--D--| |---------E----------|       |---F---|

Mb            TEAEMKASEDLKVHGIEVDTALGAILKKKGHHELEALPKAESHAKLFKI

Hb(α)         H-----GSAQVKGHGKKVADALTNAVAHVDDMPNALSALSDLHAHKLRV

Hb(β)         TPDAVMGNPKVKAHGKKVLGAFSDGLAHLNDLKGTFATLSQLHCDKLHV

              |-------G---------|      |-----------H-----------|

Mb            PIKYXEHLSXAVIHVRATKHDDEFGAPADGAMDKALELFRKDIAAKYKELGYGE

Hb(α)         DPVNFKLLSHCLLVTLAAHLPAEFTPAVHASLDKFLASVSTVLTSKWR

Hb(β)         DPENFRLLGNVLVCVLAHHFGKEFTPPVQAAYQKVVAGVANALAHKWH
```

Subsequently, the myoglobin sequence has been revised (Edmundson, 1965) and realigned (Eck and Dayhoff, 1966):

```
VLSEGEWQLVLHVWAKVEADVAGHGQDILIRLFKSHPETLEKFDRFKHLKTEAEMKASED
LKKHGVTVLTALGAILKKKGHHEAELKPLAQSHATKHKIPIKYLEFISEAIIHVLHSRHP
GNFGADAQGAMNKALELRFKDIAAKYKLEGYQG
```

These two proteins have a high content of α–helices, the locations of which are given in Table 2-1. Each chain is associated with one heme (Fig. 2-1). Heme can be physically dissociated from these proteins, and the α–helical contents are drastically reduced. Their three-dimensional structures are similar. However, the association of the two α–subunits and two β-subunits of the hemoglobin molecule is complicated, plays a vital role of the function of the entire protein, and has been extensively studied by Perutz (1998). A schematic representation of the three dimensional arrangements of these subunits in hemoglobin can be found in Perutz (1962). More detailed three dimensional structure as that shown in Fig. 2-2 can be found in a standard textbook, Lehninger Principles of Biochemistry (Nelson and Cox, 2000) and at the website <http://www.rcsb.org/pdb/>.

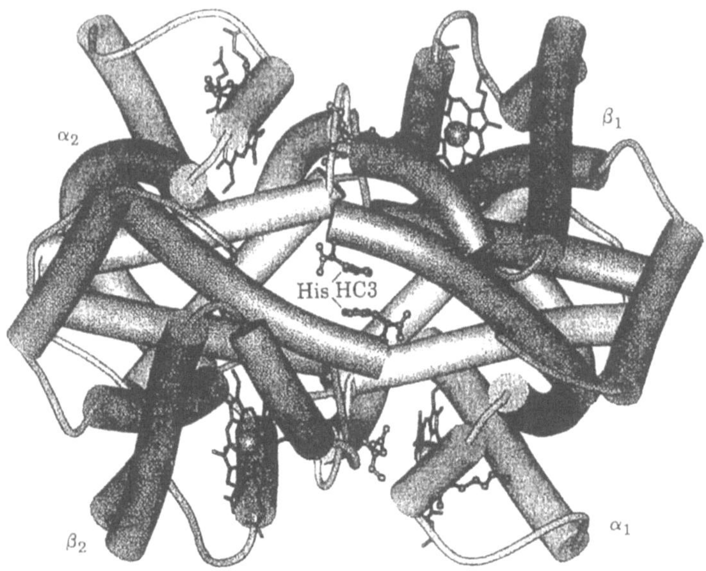

Figure 2-2. α– and β–subunits of hemoglobin from Nelson and Cox (2000) with permission.

It is important to realize that the folded structure is not static. The connecting loops or turns as well as the α-helices can move around relative to each other. The iron atom in the heme is also movable, depending on the presence or absence of oxygen molecule.

EQUILIBRIUM BINDING OF MYOGLOBIN WITH OXYGEN

The reversible binding of myoglobin with oxygen can be represented as:

$$Mb + O_2 \overset{K}{\rightleftharpoons} MbO_2,$$

where Mb is the deoxy-myoglobin, MbO_2 the oxy-myoglobin, and K the equilibrium constant. In a standard experiment, at various oxygen pressures, the relative amounts of oxy-myoglobin are measured in terms of a non-dimensional quantity, percentage saturation. It is usually represented as:

$$y = [MbO_2] / ([Mb] + [MbO_2]) .$$

Since the equilibrium constant K can be expressed as:

$$K = ([Mb] \cdot [O_2]) / [MbO_2] ,$$

We can simplify percentage saturation to:

$$y = [O_2] / (K + [O_2]) .$$

On plotting y vs. $[O_2]$, we have a curve shown in Fig. 2-3.

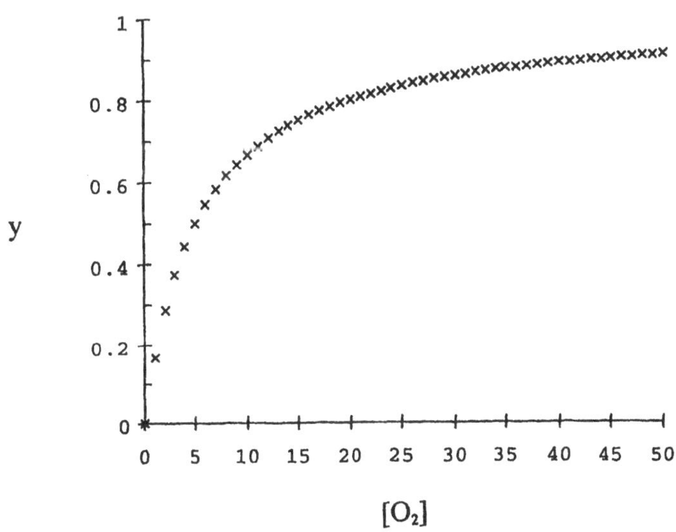

Figure 2-3. Percentage saturation, y, of myoglobin as a function of oxygen pressure, $[O_2]$.

The equilibrium constant K can be determined by drawing a horizontal line at y = 0.5 or 50%, intersecting the curve, and dropping a vertical line to the $[O_2]$ axis. It is equal to the oxygen pressure at half saturation, or $[O_2]_{1/2}$. Also, the initial slope of the curve is equal to 1/K. This same equation will be encountered again in Chapter 4 for simple enzyme kinetics.

Thus, the binding of oxygen by myoglobin is simple. However, biochemists do not like to deal with curves. Two other graphs or straight lines are commonly used. One plots 1/y as a function of $1/[O_2]$ as shown in Fig. 2-4. In this case,

$$1/y = 1 + K\, 1/[O_2] \,.$$

The straight-line extrapolation of the experimental data points insects the horizontal axis at –1/K.

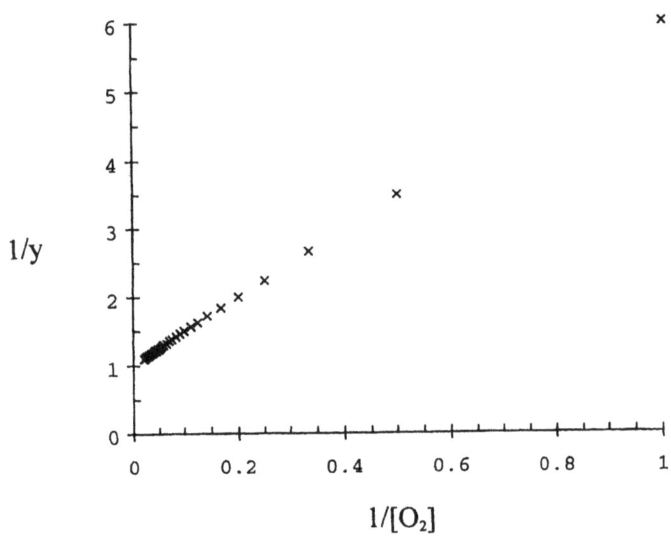

Fig. 2-4. Plot of 1/y against $1/[O_2]$.

The other plots y against $y/[O_2]$ (Fig. 2-6) or vice versa. Here,

$$y + K\, y/[O_2] = 1 \,.$$

The straight line drawn through the experimental data points intersects the horizontal axis at 1/K.

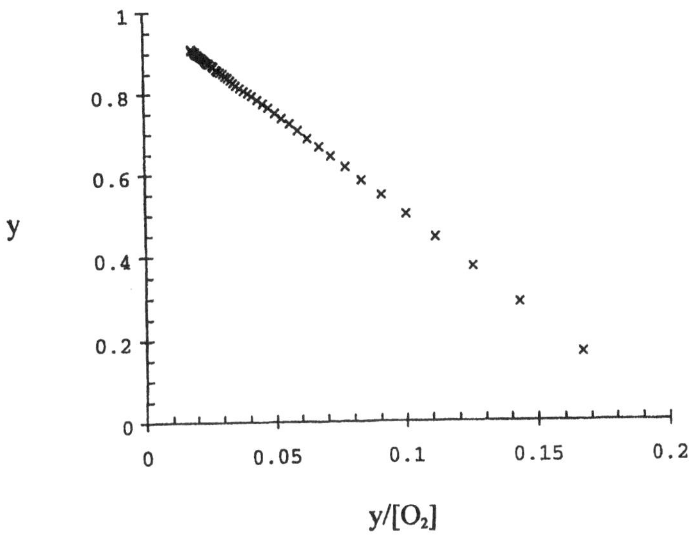

Figure 2-5. The plot of y against y/[O_2].

Experimental measurements for equilibrium saturation of myoglobin with oxygen indeed follow the theoretical curves or straight lines as shown in Figs. 2-3 to 2-5.

EXPERIMENTAL MEASUREMENTS OF SATURATION OF HEMOGLOBIN WITH OXYGEN AT EQUILIBRIUM

We can easily realize from the saturation curve shown in Fig. 2-2 that myoglobin would not be a good oxygen carrier protein from our lungs to our tissues, since the oxygen pressure in our tissues is not near zero. We need a protein having a steep slope between the oxygen pressures of our tissues and our lungs. Indeed, hemoglobin has evolved just for that purpose. In order to understand its properties, careful experimental measurements are essential. The best study was by Roughton *et al.* in 1955. Their results, together with experimental error, are listed in Table 2-2.

Table 2-2. Experimental measurements of hemoglobin saturation with errors (Roughton *et al.*, 1955).

Oxygen pressure (mm Hg)	Percentage saturation (error)
0.306	0.782 (0.05)
0.408	1.023 (0.05)
2.14	11.17 (0.5)
2.943	24.51 (0.5)
3.141	27.68 (0.5)
5.574	68.78 (0.5)
7.349	82.79 (0.5)
10.220	92.72 (0.5)
29.5	98.85 (0.05)
44.2	99.23 (0.05)

One may wonder why they were so meticulous about their errors. Indeed, if percentage saturation, y, is plotted against oxygen pressure, $[O_2\}$, as shown in Fig. 2-6, the error bars are not visible.

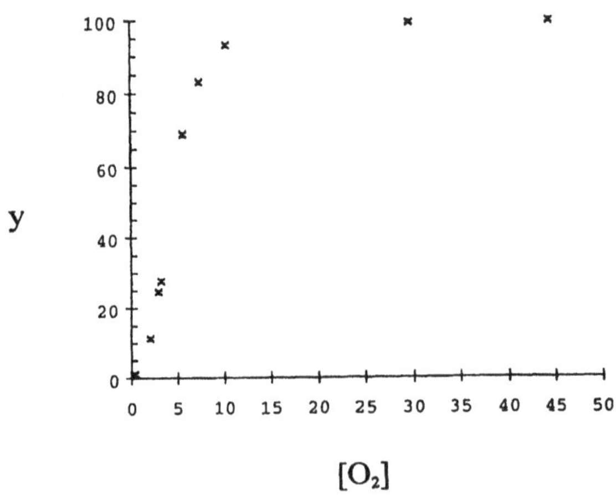

Figure 2-6. Percentage saturation of hemoglobin, y, against oxygen pressure, $[O_2]$.

However, it is clear that the saturation curve is "S" shaped. The location of the steep slope depends on the pH of the solution, known as the Bohr effect, some small molecules, e.g. 2,3-diphosphoglycerate, etc. Over the years, numerous theoretical models have been proposed to explain this interesting behavior of hemoglobin. Three of them will be discussed here.

ONE-CONSTANT MODEL

The simplest theoretical model is that proposed by Hill (1910), as an extension to the oxygen saturation of myoglobin discussed above, i.e.:

$$Hb + 4\,O_2 \overset{K}{\rightleftharpoons} Hb([O_2])_4\,,$$

since hemoglobin has two α–subunits and two β–subunits. He realized that this was not possible, since a five molecular collision would be required. However, one can carry out the formality and derive an equation for the percentage saturation, y, just as in the case of myoglobin. We get:

$$y = [O_2]^4 / (K + [O_2]^4)\,.$$

Indeed, this equation can fit the experimental data plotted in Fig. 2-6 very well. At very low oxygen pressure, y will increase as the fourth power of $[O_2]$. At higher oxygen pressure, y will approach 1, and can exhibit the property of an "S" shaped curve. The equilibrium constant K can also be determined as before, and is equal to $[O_2]_{1/2}^4$.

Hill proposed that instead of the y vs. $[O_2]$ plot, a more sensitive plot would be needed. That plot is now known as the Hill's plot. Instead of y, he calculated:

$$y/(1\text{-}y) = [O_2]^4 / K$$

$$= [O_2]^4 / [O_2]_{1/2}^4$$

$$= ([O_2] / [O_2]_{1/2})^4 .$$

Therefore, Hill proposed to plot $\ln\{y/(1-y)\}$ against $\ln\{[O_2] / [O_2]_{1/2}\}$. In this case, his one-constant model of hemoglobin saturation would give a straight line of slope of four, since

$$\ln\{y/(1-y)\} = 4 \ln\{[O_2] / [O_2]_{1/2}\} .$$

This is the well known Hill's plot.

To compare with the experiment results of Roughton *et al.* (1955) in Table 2-2, we need to estimate the value of $[O_2]_{1/2}$ from the plot in Fig. 2-6. Using the two experimental measurements closest to 50% saturation, we have:

$$\frac{68.78 - 27.68}{5{,}574 - 3.141} = \frac{50.00 - 27.68}{[O_2]_{1/2} - 3.141} .$$

Therefore, $[O_2]_{1/2}$ is around 4.46 mm Hg.

The experimental Hill's plot is shown in Fig. 2-7, including error bars of 0.5% for all ten experimental measurements of percentage saturation. It is immediately obvious that this error renders the measurements at very low and very high percentage saturation useless. For this reason, Roughton *et al.* were extremely careful of reducing the errors for these measurements down to 0.05%.

The slope of the experimental result around 50% saturation is usually between 2.8 and 3.1, indicating clearly that the one-constant model of hemoglobin saturation is not correct. In fact, at one time, someone even suggested that hemoglobin might have only three subunits. The three-dimensional structure determined by X-ray diffraction studies obviously eliminated that possibility (Fig. 2-2). Many models based on two molecular collisions, i.e. sequential binding of oxygen molecules to hemoglobin, have thus been proposed subseqeuntly.

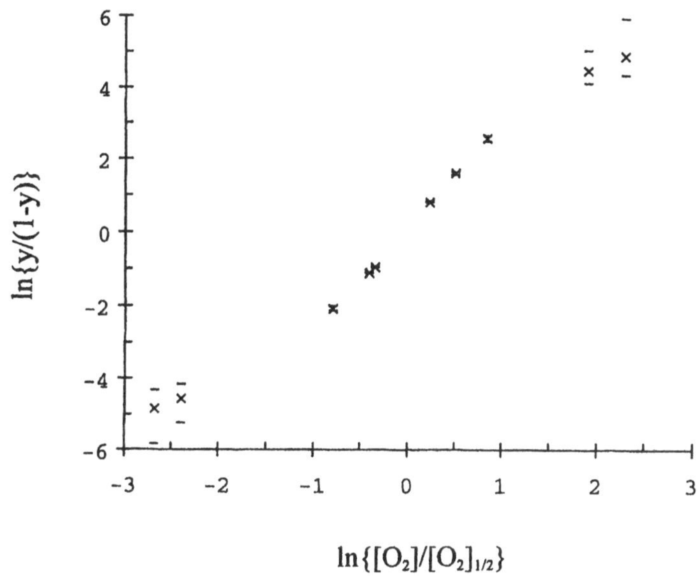

Figure 2.7. Hill's plot for the experimental results of Roughton *et al.* with error bars of 0.5% for all ten measurements.

TWO-CONSTANT MODEL

Pauling proposed two theoretical models for hemoglobin saturation in 1935, both with two constants. Only his tetrahedral model will be discussed here. In this case, he assumed that the two α–subunits and two β–subunits are arranged in a tetrahedral configuration, as verified by X-ray diffraction studies years later. Different numbers of oxygen molecules, zero to four, would bind to these four subunits. On the addition of an oxygen molecule to the heme, he assumed that there would be a free energy change of RT ln K' which defined the constant K', where R is a universal constant and T the absolute temperature. He further assumed that the stabilizing energy associated with two neighboring oxygenated hemes could be denoted by RT ln α which defined the constant α. If we denote the relative amount of deoxy-hemoglobin as 1, then the relative amounts of hemoglobin with one to four oxygen molecules would be:

one oxygen molecule: $4\,K'\,[O_2]$,

two oxygen molecules: $6\,\alpha\,K'^2\,[O_2]^2$.

three oxygen molecules: $4\,\alpha^3\,K'^3\,[O_2]^3$, and

four oxygen molecules: $\alpha^6\,K'^4\,[O_2]^4$.

The factors 4, 6 and 4 denote possible alternative associations of oxygen with the subunits of hemoglobin. The powers of α indicate the numbers of nearest neighbor connections in the tetrahedral model (Fig. 2-8).

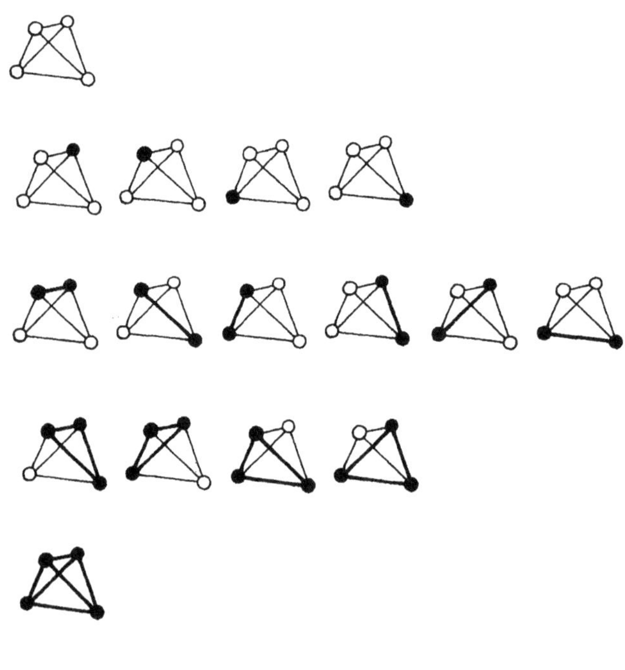

Figure 2-8. All possible states in the tetrahedral model of Pauling (1935).

In order to calculate percentage saturation, y, at a certain oxygen pressure $[O_2]$, we need to divide the total number of oxygen molecules associated with hemoglobin by the total number of available sites for oxygen molecules:

$$y = \frac{4\,K'\,[O_2] + 12\,\alpha\,K'^2\,[O_2]^2 + 12\,\alpha^3\,K'^3\,[O_2]^3 + 4\,\alpha^6\,K'^4\,[O_2]^4}{4\,\{1 + 4\,K'\,[O_2] + 6\,\alpha\,K'^2\,[O_2]^2 + 4\,\alpha^3\,K'^3\,[O_2]^3 + \alpha^6\,K'^4\,[O_2]^4\}}\,.$$

A factor of 4 can be cancelled from the numerator and the denominator, giving the equation in Pauling's original paper. He used p to denote $[O_2]$.

To compare his theoretical model with the experimental data of Roughton *et al.* in the Hill's plot, we can estimate the asymptotic expressions of y/(1-y) for very low and very high oxygen pressures. Since

$$\frac{y}{1-y} = \frac{K'\,[O_2] + 3\,\alpha\,K'^2\,[O_2]^2 + 3\,\alpha^3\,K'^3\,[O_2]^3 + \alpha^6\,K'^4\,[O_2]^4}{1 + 3\,K'\,[O_2] + 3\,\alpha\,K'^2\,[O_2]^2 + \alpha^3\,K'^3\,[O_2]^3}\,,$$

we have, for very low oxygen pressure, the following:

$$\ln\{y/(1-y)\} = \ln\{[O_2]\,/\,[O_2]_{1/2}\} + \ln\{K'\,[O_2]_{1/2}\}\,,$$

which is a straight line of slope one in the Hill's plot. Since $[O_2]_{1/2} = 4.46$ mm Hg, K' can thus be determined from the experimental data of Roughton *et al.* Similarly, for very high oxygen pressure, we have:

$$\ln\{y/(1-y)\} = \ln\{[O_2]\,/\,[O_2]_{1/2}\} + \ln\{\alpha^3\,K'\,[O_2]_{1/2}\}\,,$$

which is also a straight line of slope one in the Hill's plot. The value of α can also be determined experimentally. Indeed, very few experimental measurements of hemoglobin saturation other than those of Roughton *et al.* were accurate enough to exhibit these important features in the Hill's plot.

Pauling's theoretical model has a very important hidden finding. To satisfy the requirement that at $[O_2] = [O_2]_{1/2}$, $y = ½$, the equation for y/(1-y) gives the following:

$$1 + 2\,K'\,[O_2]_{1/2} = 2\,\alpha^3\,K'^3\,[O_2]_{1/2}^3 + \alpha^6\,K'^4\,[O_2]_{1/2}^4\,.$$

This equation can be considered as a quadratic equation for α^3. Since α has to be positive by definition, α^3 has to be positive also. The two solutions for α^3 are:

$$\alpha^3 = \frac{-\,2\,K'^3\,[O_2]_{1/2}^3 + \{4\,K'^6\,[O_2]_{1/2}^6 + 8\,K'^5\,[O_2]_{1/2}^5 + 4\,K'^4\,[O_2]_{1/2}^4\}^{1/2}}{2\,K'^4\,[O_2]_{1/2}^4},$$

$$= \frac{1}{K'^2\,[O_2]_{1/2}^2},$$

$$\text{or } \alpha^3 = \frac{-\,2\,K'^3\,[O_2]_{1/2}^3 - \{4\,K'^6\,[O_2]_{1/2}^6 + 8\,K'^5\,[O_2]_{1/2}^5 + 4\,K'^4\,[O_2]_{1/2}^4\}^{1/2}}{2\,K'^4\,[O_2]_{1/2}^4}.$$

$$= \frac{-\,1 - 2\,K'\,[O_2]_{1/2}}{K'^2\,[O_2]_{1/2}^2}.$$

The second solution is not valid, since it is negative. The first solution gives:

$$K'\,[O_2]_{1/2} = 1\,/\,\{\alpha^3\,K'\,[O_2]_{1/2}\}\,,$$

or

$$\ln\,\{K'\,[O_2]_{1/2}\} = -\,\ln\,\{\alpha^3\,K'\,[O_2]_{1/2}\}\,.$$

In other words, the Hill's plot should be symmetric around the origin. The experimental measurements of Roughton *et al.* gave:

$$\ln \{K' [O_2]_{1/2}\} = -2.173,$$

and

$$\ln \{\alpha^3 K' [O_2]_{1/2}\} = 2.565,$$

which are reasonably close to the above requirement. For this interesting reason, Pauling's model has been "re-discovered" after more than 60 years. Using the average of these two numbers, we get K' = 0.021 and α = 4.85 which are reasonably close to what Pauling got originally.

THREE-CONSTANT MODEL

The most commonly quoted theoretical model in biochemistry textbooks for equilibrium hemoglobin saturation with oxygen is the MWC model proposed by Monod *et al.* (1965). They made the following assumptions:

1. The α–subunits and β–subunits are the same.

2. Hemoglobin molecules exist in two different states: T for tense (or taut) and R for relaxed. These two states are in equilibrium.

3. Oxygen molecules bind to the hemoglobin subunits sequentially. The equilibrium constants are the same, one for the T–state and another for the R–state.

4. At low oxygen pressure, hemoglobin is in the T–state. Transition occurs from the T–state to the R–state as the oxygen pressure is increase.

The T- and R-states of hemoglobin are designated as T_0, T_1, T_2, T_3, T_4 and R_0, R_1, R_2, R_3, R_4 respectively with the subscripts indicating the number of oxygen molecules bound to that state. Thus,

$$T_0 + O_2 \xrightleftharpoons{k_T} T_1 ,$$

where k_T is the equilibrium constant given by:

$$k_T = \frac{[T_0]\,[O_2]}{[T_1]} \frac{4}{1} .$$

The factor 4 in the last equation is required to take into consideration of the four identical subunits capable of binding the first oxygen molecule, while the factor 1 indicates that there is only one subunit with a bound oxygen molecule. We also have, for subsequent bindings of oxygen molecules:

$$k_T = \frac{[T_1]\,[O_2]}{[T_2]} \frac{3}{2} = \frac{[T_2]\,[O_2]}{[T_3]} \frac{2}{3} = \frac{[T_3]\,[O_2]}{[T_4]} \frac{1}{4} .$$

For the R-states, the equilibrium constant k_R is similarly defined. The equilibrium constant between the T- and R-states is

$$K = [T_0] / [R_0] .$$

Percentage saturation, y, can again be defined by counting the number of oxygen molecules on each of the states divided by the total number of available oxygen binding sites of all the states. It is, with the use of the summation notation,:

$$y = \frac{\Sigma\, i\, [T_i] + \Sigma\, i\, [R_i]}{4\{\Sigma\, [T_i] + \Sigma\, [R_i]\}} ,$$

where the summation extends over all subscripts, i.e. i = 0 to 5. The right hand side of the equation can be expressed as a function of $[O_2]$ with the three constants, k_T , k_R , and K. To obtain that expression, we need to calculate the above four sums.

$$\Sigma\, i\, [T_i] = [T_1] + 2\,[T_2] + 3\,[T_3] + 4\,[T_4]$$

$$= 4\,[T_0]\,([O_2]/k_T) + 12\,[T_0]\,([O_2]/k_T)^2 + 12\,[T_0]\,([O_2]/k_T)^3$$

$$+ 3\,[T_0]\,([O_2]/k_T)^4$$

$$= 4\,[T_0]\,([O_2]/k_T)\,\{1 + ([O_2]/k_T)\}^3\,.$$

Similarly, we can derive the following results:

$$\Sigma\, i\, [R_i] = 4\,[R_0]\,([O_2]/k_R)\,\{1 + ([O_2]/k_R)\}^3\,,$$

$$\Sigma\,[T_i] = [T_0]\,\{1 + \{[O_2]/k_T)\}^4\,,$$

and $$\Sigma\,[R_i] = [R_0]\,\{1 + ([O_2]/k_R\}^4\,.$$

With these simplifications, we get:

$$y = \frac{K\,\{[O_2]/k_T)\,\{1 + ([O_2]/k_T\}^3 + ([O_2]/k_R)\,\{1 + ()O_2]/k_R\}^3}{K\,\{1 + ([O_2]/k_T)\}^4 + \{1 + ([O_2]/k_R)\}^4}\,,$$

with the three equilibrium constants clearly indicated. In the original paper, Monod *et al.* (1965) made the following substitutions:

$$\alpha = [O_2]/k_R ,$$

and

$$c = k_R/k_T .$$

Therefore, percentage saturation satisfies the following equation:

$$y = \frac{Kc\alpha \{1 + c\alpha\}^3 + \alpha \{1 + \alpha\}^3}{K \{1 + c\alpha\}^4 + \{1 + \alpha\}^4} .$$

A casual reader would consider this as a two-constant model having K and c as equilibrium constants. However, α is not $[O_2]$ and a third constant is hidden in α. Comparing to some experimental results, Monod *et al.* found that K was around 10^4 and c around 10^{-2}.

To estimate the values of k_T . k_R , and K of the MWC model based on the experimental measurements of Roughton *et al.* (1955), we can again use the very low and very high oxygen pressure approximations of y/(1-y). In this case, we have for very low oxygen pressure,

$$\ln \{y/(1-y)\} = \ln \{[O_2] / [O_2]_{1/2}\} + \ln \{[O_2]_{1/2}/k_T\} ,$$

and for very high oxygen pressure,

$$\ln \{y/(1-y)\} = \ln \{[O_2] / [O_2]_{1/2}\} + \ln \{[O_2]_{1/2}/k_R\} ,$$

using the approximate values of K and c determined by Monod *et al.* (1965). The value of K can be determined using the requirement of $[O_2] = [O_2]_{1/2}$, y = ½, together with the values of k_T and k_R . Thus, we get initially:

$k_T = 39.2$ mm Hg,

$k_R = 0.343$ mm Hg,

and $K = 2.69 \times 10^4$.

The value of c is then about 8.57×10^{-3}, in reasonably good agreement with what Monod *et al.*(1965) found originally. The value of K, however, is quite different. More accurate values of these equilibrium constants can be obtained by least square fitting of the theoretical equation to the experimental data as discussed below.

LEAST SQUARE FITTING

The three models discussed above can all be fitted to experimental measured data of hemoglobin saturation plotted against oxygen pressure, i.e. the y vs. $[O_2]$ plot. However, for the same data displayed by the Hill's plot, the one-constant model does not fit at all. The two-constant model can be fitted reasonably closely, but not completely, due to the symmetry property of the model. On the other hand, the three-constant model may be fitted quite closely. But, what is "fitting"? To define "fitting" precisely, we need to introduce the idea of least square.

In our present discussion, for any experimentally measured oxygen pressure, $[O_2]$, the corresponding experimentally determined percentage saturation, y, is converted to $\ln \{y/(1-y)\}$, to give an experimental data point on the Hill's plot. On the other hand, we can substitute the experimentally measured value of $[O_2]$ into the equation derived from the MWC model:

$$\frac{y}{1-y} = \frac{K([O_2]/k_T)\{1+([O_2]/k_T)\}^3 + ([O_2]/k_R)\{1+([O_2]/k_R)\}^3}{K\{1+([O_2]/k_T)\}^3 + \{1+([O_2]/k_R)\}^3},$$

to obtain a theoretical value of $\ln \{y/(1-y)\}$ which depends on the three equilibrium constants. The difference between the experimental and theoretical values of ln {y/(1-y)} is then squared. For the ten measurements by Roughton *et al.*, these ten squares are summed to give $S(k_T,k_R,K)$. This sum of squares is then minimized with respect to the three equilibrium constants. These three constants can be varied either independently or together until S reaches a minimum.

Starting with the values of the three constants listed before, we get a value of S = 0.13715. In the Hill's plot, we can look at the discrepancies between experimental points and theoretical values of ln {y/(1-y)}. They are listed in Table 2-3 with these initial values of k_T , k_R and K for the ten experimentally measure oxygen pressure.

Table 2-3. Experimental and theoretical values of ln {y/(1-y)} with k_T = 39.2 mm Hg, K_R = 0.343 mm Hg and K = 2.69 X 10^4.

PO_2 (mm Hg)	experimental	theoretical
0.306	-4.843	-4.825
0.408	-4.572	-4.523
2.014	-2.073	-2.197
2.943	-1.125	-1.227
3.141	-0.960	-1.043
5.574	0.790	0.679
7.239	1.571	1.491
10.220	2.544	2.365
29.5	4.454	4.258
44.2	4.859	4.748

The results in Table 2-3 clearly indicate that the initial estimations of the three constants give a good fit for experimental data points at very low oxygen pressure, and not so good for the rest. Since the equation for ln {y/(1-y)} is highly non-linear, adjusting the three constants can sometimes get to a local minimum of the sum of squares, S. Therefore, it becomes essential to plot the experimental and theoretical values together to examine their differences.

By adjusting these constants in a sufficiently broad range, the values of S for different sets are listed in Table 2-4:

Table 2-4. Values of $S(k_T,k_R,K)$ for different sets of k_T, k_R and K.

k_T	k_R	K	$S(k_T,k_R,K)$
39.2 mm Hg	0.343 mm Hg	2.69×10^4	0.13715
41.0 mm Hg	0.341 mm Hg	2.28×10^4	0.04708
40.5 mm Hg	0.315 mm Hg	3.16×10^4	0.02170
40.0 mm Hg	0.300 mm Hg	3.90×10^4	0.01574
40.0 mm Hg	0.300 mm Hg	3.92×10^4	0.01563
39.9 mm Hg	0.300 mm Hg	3.92×10^4	0.01561

The value of S may be reduced even further. Table 2-4 indicates that the initial estimation for K is not very accurate, the reason being the inaccurate value of $[O_2]_{1/2}$ obtained from the y against $[O_2]$ plot (Fig. 2-6). Calculating backwards for the last set of k_T, k_R and K, we get:

$$[O_2]_{1/2} = 4.29 \text{ mm Hg} .$$

With these new values, a revised Hill's plot is shown in Fig. 2-9 with the actual error bars measured by Roughton *et al.* (1955), and the theoretically calculate values of ln $\{y/(1-y)\}$ based on the MWC model and the best set of three constants. The fit is reasonable, but still not perfect.

The ratio c (= k_R/k_T) changes to 0.752×10^{-3} only slight different from the original estimation of Monod *et al.*(1965). However, the value of K is quite different, being about four times larger.

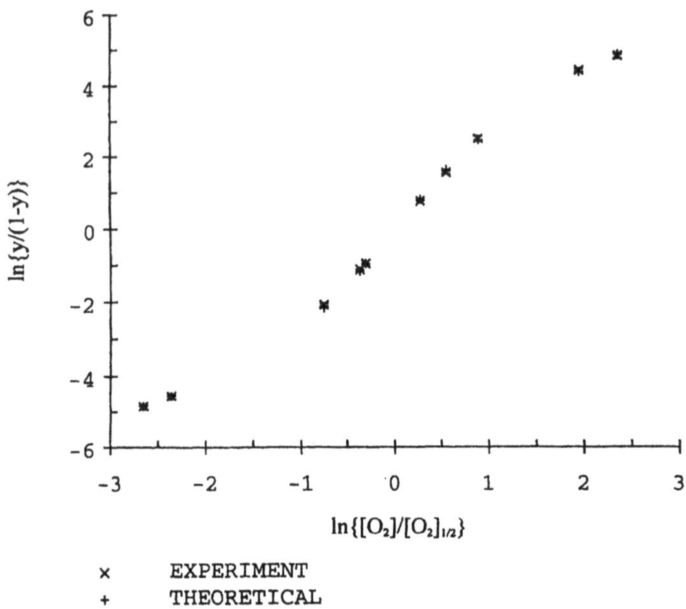

Figure 2-9. Hill's plot for the experimental measurements of Roughton *et al.* and the theoretical values based on the MWC model.

PREDICTION BASED ON THESE CONSTANTS

The very large value of K indicates that at low oxygen pressure, hemoglobin molecules are mostly in the T-state. Therefore, $[R_0]$ can be neglected completely with respect to $[T_0]$. The ratio of $[T_1] / [R_1]$ is around 2.95 X 10^2, so that $[R_1]$ can also be neglected with respect to $[T_1]$. On the other hand, the ratio of $[T_2] / [R_2]$ is around 12.22, i.e. they are of similar concentrations. At high oxygen pressure, $[T_3] / [R_3] = 1.67 \text{ X } 10^{-2}$, and $[T_4] / [R_4] = 1.25 \text{ X } 10^{-4}$. Therefore, $[T_3]$ and $[T_4]$ can be neglected. In short, according to the MWC model, as the oxygen pressure increases, roughly around 50% saturation, the T–state changes into the R–state (Fig. 2-10). This transition is the most important prediction of the MWC model (Monod *et al.,* 1955).

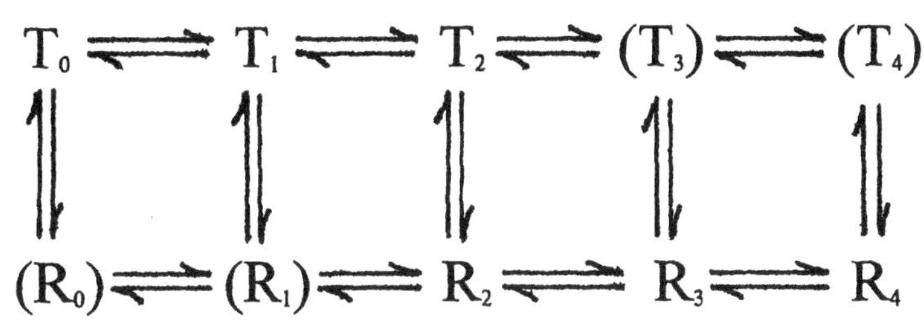

Figure 2-10. Various states in the MDC model.

FURTHER EXPERIMENTAL STUDIES

Perutz *et al.* (1998) reviewed the current status of this problem and gave a detailed account of how the transition occurred between the T- and R-states (Fig. 2-11).

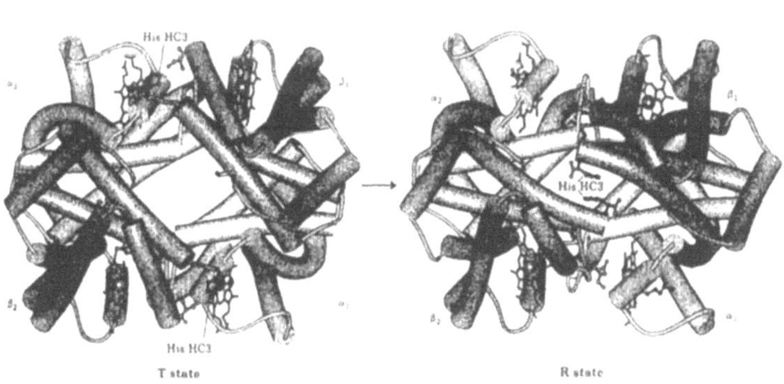

Figure 2-11. The T to R transition (from Nelson and Cox, 2000, with permission).

There are relative movements between the α– and β–subunits, as well structural changes involving salt-bridges, α-helices, iron atoms, etc. Many of the recent experimental findings point to the idea that oxygen molecules would initially bind to the α–subunits first. When both α–subunits contain oxygen molecules, the subsequent configuration changes would then allow β–subunits to bind oxygen molecules. In other words, we have:

$T_0 \qquad [\,\alpha\,\alpha\,\beta\,\beta\,]_T$

$T_1 \qquad [\,\underline{\alpha}\,\alpha\,\beta\,\beta\,]_T$

$T_2 \qquad [\,\underline{\alpha}\,\underline{\alpha}\,\beta\,\beta\,]_T$

$R_2 \qquad [\,\underline{\alpha}\,\underline{\alpha}\,\beta\,\beta\,]_R$

$R_3 \qquad [\,\underline{\alpha}\,\underline{\alpha}\,\underline{\beta}\,\beta\,]_R$

$R_4 \qquad [\,\underline{\alpha}\,\underline{\alpha}\,\underline{\beta}\,\underline{\beta}\,]_R$

where the underlines indicate the presence of bound oxygen molecules on the subunits. Thus, the assumption in the MWC model of α–subunits and β–subunits being identical has to be modified. Consequently, the equations for the equilibrium constants become:

$$k_T = \frac{[T_0]\,[O_2]}{[T_1]}\,\frac{2}{1} = \frac{[T_1]\,[O_2]}{[T_2]}\,\frac{1}{2},$$

$$k = [T_2] \,/\, [R_2]\,,$$

and $$k_R = \frac{[R_2][O_2]}{[R_3]} \frac{2}{1} = \frac{[R_3][O_2]}{[R_4]} \frac{1}{2} .$$

Similar modifications will be required for the expression of percentage saturation:

$$y = \frac{[T_1] + 2[T_2] + 2[R_2] + 3[R_3] + 4[R_4]}{4\{[T_0] + [T_1] + [T_2] + [R_2] + [R_3] + [R_4]\}} .$$

From this, we can derive the following:

$$\frac{y}{1\text{-}y} = \frac{k\{(k_T/[O_2]) + 1\} + \{1 + 3([O_2]/k_R) + 2([O_2]/k_R)^2\}}{k\{2(k_T/[O_2])^2 + 3(k_T/[O_2]) + 1\} + \{1 + ([O_2]/k_R)\}} .$$

On using the experimental data of Roughton *et al.* (1955) at very low and very high oxygen pressures, we note that the value of k_T is reduced by a factor of two, while that of k_R increased by a factor of two, i.e.

$$k_T = 19.6 \text{ mm Hg},$$

and $$k_R = 0.686 \text{ mm Hg}.$$

The requirement of $[O_2] = [O_2]_{1/2}$ at $y = ½$ gives:

$$k = 2.06,$$

which is equal to the value of $K(k_R/k_T)^2$ in the MWC model. Again, by least square fitting to the entire set of experimental measurements of Roughton *et al.* (1955) in the Hill's plot, these three equilibrium constants will be changed slightly.

OTHER STUDIES ON HEMOGLOBIN

We have only reviewed three of the relatively simple models on equilibrium saturation of hemoglobin with oxygen. There are many other models, e.g. Adair's (1925), Koshland's (1966). etc., proposed during the past century. In addition, this protein also transports carbon dioxide, and has many additional interesting properties.

Over the years, numerous studies have been carried out on its kinetic properties, replacement of iron by other transition elements, effects of pH, binding of small molecules, genetic mutants, etc., just to name a few. A close collaboration between experimental measurements and theoretical models plays vital roles in all of these studies. There are also many diseases associated with hemoglobin abnormalities, e.g. Hb Kansas, Hb Yakima, Hb Zurich, Hb S, Hb Lepore, α and β thalassemia, persistence of fetal γ chain, etc. Only recently, specified amino acid residue changes have been used to study the detailed molecular mechanisms of hemoglobin functions (Bettati *el al.*, 1998; Tsai *et al.*, 2000).

2,3-DIPHOSPHOGLYCERATE

This small molecule is highly concentrated inside our red blood cells. Each hemoglobin molecule binds one molecule of 2,3-diphosphoglycerate. Its presence is essential to the S-shaped hemoglobin saturation curve. Perutz (1970) has reasoned that it binds tightly to the β–subunits of deoxy-hemoglobin or the T–state. On transition to oxy-hemoglobin or the R–state, 2,3-diphosphoglycerate drops out. Whether it binds to the two β–subunits symmetrically (Arnone, 1972) or asymmetrically (Pomponi *et al.*, 2000) remains uncertain.

OTHER ALLOSTERIC PROTEINS AND ENZYMES

Conformational changes of proteins due to the binding of small molecules occur frequently in various biochemical reactions. Since the first description of this phenomenon, known as allosteric regulation, in 1963 by Monod *et al.*, numerous papers are published yearly about various proteins and enzymes exhibiting this behavior. Hemoglobin is just one of the proteins which have been carefully analyzed. An enzyme, aspartate transcarbamylase (Gerhart and Schachman, 1968), has also been extensively studied.

To find other examples, the website <http://www.ncbi.nlm.nih.gov> can be used to search for "allosteric enzymes" under PubMed. Several thousand papers will be listed. However, careful measurements such as those of Roughton *et al.* (1955) are rare. Without such experimental data, it would be difficult to decide which models are more close to reality, as we have clearly illustrated in this Chapter.

REFERENCES

Adair GS (1925) The hemoglobin system. VI: The oxygen dissociation curve of hemoglobin. *J. Biol. Chem.*, **63**, 529-545.

Arnone A (1972) X-ray diffraction study of binding of 2,3-diphosphoglycerate to human deoxyhemoglobin. *Nature*, **237**, 146-149.

Battati S, Morrarelli A and Perutz MF (1998) Allosteric mechanism of hemoglobin: rupture of salt-bridges raises the oxygen affinity of the T-structure. *J. Mol. Biol.*, **281**, 581-585.

Eck RV and Dayhoff MO (1966) *Atlas of Protein Sequence and Structure*. National Biomedical Research Foundation, Silver Springs, MD.

Edmundson AB (1965) Amino-acid sequence of sperm whale myoglobin. *Nature*, **205**, 883-887.

Gerhart JC and Schachman HK (1968) Allosteric interactions in aspartate transcarbamylase. II. Evidence for different conformational states of the protein in the presence and absence of specific ligands. *Biochemistry*, **7**, 538-552.

Hill AV (1910) The possible effects of the aggregation of the molecules of hemoglobin on its dissociation curves. *J. Physiol.*, **40**, iv-vii.

Koshland DE Jr, Nemethy G and Filmer D (1966) Comparison of experimental binding data and theoretical models in proteins containing subunits. *Biochemistry*, **5**, 365-385.

Monod J. Changeux J-P and Jacob F (1963) Allosteric proteins and cellular control systems. *J. Mol. Biol.*, **6**, 306-329.

Monod J, Wyman J and Changeux J-P (1965) On the nature of allosteric transitions: a plausible model. *J. Mol. Biol.*, **12**, 88-118.

Nelson DL and Cox MM (2000) *Lehninger Principles of Biochemistry, 3rd Edition*. Worth Publishers, New York.

Nobbs CL, Watson HC and Kendrew JC (1966) Structure of deoxymyoglobin: a crystallographic study. *Nature*, **209**, 339-341.

Pauling L (1935) The oxygen equilibrium and its structural interpretation. *Proc. Nat. Acad. Sci. USA*, **21**, 186-191.

Perutz MF (1962) *Proteins and Nucleic Acids*. Elsevier Publishing Company, New York.

Perutz MF (1970) Stereochemistry of cooperative effects in hemoglobin. *Nature*, **228**, 726-739.

Perutz MF, Wilkinson AJ, Paoli M and Dodson GG (1998) The stereochemical mechanism of cooperative effects in hemoglobin revisited. *Annu. Rev. Biophys. Biomol. Struct.*, **27**, 1-34.

Pomponi M, Bertonati C, Fuglei E, Wiig O and Derocher AE (2000) 2,3-DPG-Hb complex: a hypothesis for an asymmetric binding. *Biophys. Chem.*, **84**, 253-260.

Roughton FJW, Otis AB and Lyster RLJ (1955) The determination of the individual equilibrium constants of the four intermediate reactions between oxygen and sheep hemoglobin. *Proc. Roy. Soc. (London)*, **B144**, 29-54.

Tsai CH, Fang TY, Ho NT and Ho C (2000) Novel recombinant hemoglobin, rHb (betaN108Q), with low oxygen affinity, high cooperativity, and stable against autooxidation. *Biochemistry*, **39**, 13719-13729.

EXERCISE

Determine the three equilibrium constants in the Perutz's modification of the MWC model by least square fitting to the experimental measurements of Roughton *et al.* in the Hill's plot. Draw the curve to visualize that it indeed passes through all the ten experimental data points very closely. The value of $[O_2]_{1/2}$ may be adjusted slightly to make sure that the interpolated straight line of the middle experimental data points passes through the origin.

CHAPTER 3

CO-TRANSDUCTION AND LOCATION OF *Escherichia coli* GENETIC MARKERS

INTRODUCTION

Escherichia coli has been one of the central tools to study the basic principles of biochemistry, molecular biology and cell biology. Non-pathogenic strains were used in many laboratories. Metabolic pathways have been delineated carefully. Mutants were isolated with certain selected defective enzymes involved in these pathways. Various methods of genetic manipulation using these mutants have been developed. The collection of these experimental findings verified the circular chromosomal structure. Then, a detailed *E. coli* genetic map was constructed, followed by complete sequencing of its genomic DNA.

In this chapter, one of the genetic methods developed for *E. coli*, transduction (Lederberg *et al.*, 1951; Zinder and Lederberg, 1952; Lennox, 1955), will be discussed in detail. Experimentally, two genetic markers can be co-transduced if they are sufficiently close to each other. The frequency of co-transduction will be related to the distance between these two markers. This relationship can be derived theoretically by considering all aspects of the biological process of transduction. Many experiments were also carried out with three nearby genetic markers, known as three-point analysis. Some researchers even studied four closely linked markers.

In order for an experimental study of transduction to be successful, it is essential to determine the approximate locations of the involved genetic markers. Such estimations can be achieved by another genetic manipulation also developed for *E. coli* known as conjugation (Lederberg and Tatum, 1946). This process involves the transfer of the entire *E. coli* chromosome from the donor to the recipient in 100 minutes at 37°C in a rich medium, and will thus be reviewed first.

CONJUGATION

Strain *E. coli* K-12 can carry an F plasmid capable of integrating into the *E. coli* chromosome at various locations in different orientations. They can transfer their DNA during synthesis, at high frequency into a recipient *E. coli* through physical contact, and thus designated as Hfr or high frequency of recombination. The donor-recipient pair can be broken artificially by high shear rate in a blender, for example, and the transfer of genetic material is thus interrupted.

A typical experiment consisted of mixing a donor strain with a good gene and a recipient strain with a defective gene. For example, the recipient might have a defective mutation of one of the genes for the synthesis of amino acid threonine, designated as thr^-. Without providing threonine in the growth medium, the recipient strain could not grow. On the other hand, the donor carried the normal gene, designated as thr^+. In order to select recombinants with thr^+ gene, no threonine was added to the growth medium in which the mixture was plated. However, another genetic marker was required to prevent the growth of the donor strain. Usually, the gene which provided resistant to streptomycin, $strM^+$, was used. The recipient was $strM^+$ while the donor was $strM^-$. The addition of streptomycin to the medium without threonine could only support the growth of the desired recombinant.

At various times after mixing the donor and recipient strains together, the culture was vigorously blended for a minute before plating. If the thr^+ gene from the donor strain had not been transferred into the recipient strain, no thr^+ recombinants could result, except by reversion of thr^- to thr^+ usually at a very low frequency around 10^{-7}. Therefore, no colony would appear on the selecting plate. After a certain duration, the thr^+ gene from the donor would enter the recipient. By recombination with at least two cross-overs, the thr^+ gene could replace the thr^- gene on the recipient chromosome, to give rise to the required recombinants. They would form colonies on the selecting plates. The number of recombinants was plotted against time after mixing. The intersection of this curve with the time-axis was recorded. That time would be proportional to the length of the *E. coli* chromosome from the point of F plasmid integration to the genetic marker *thr*.

Using different Hfr's of different orientations and starting points of transfer in such interrupted mating experiments, many pioneer researchers in this field have established the approximate locations of numerous genetic markers on the *E. coli* chromosome map. The chromosome is circular and has a length of 100 minutes, i.e. the time required to transfer the entire chromosome under ideal conditions (Fig. 3-1). Position 0 minute designates the location of the genetic marker *thr*.

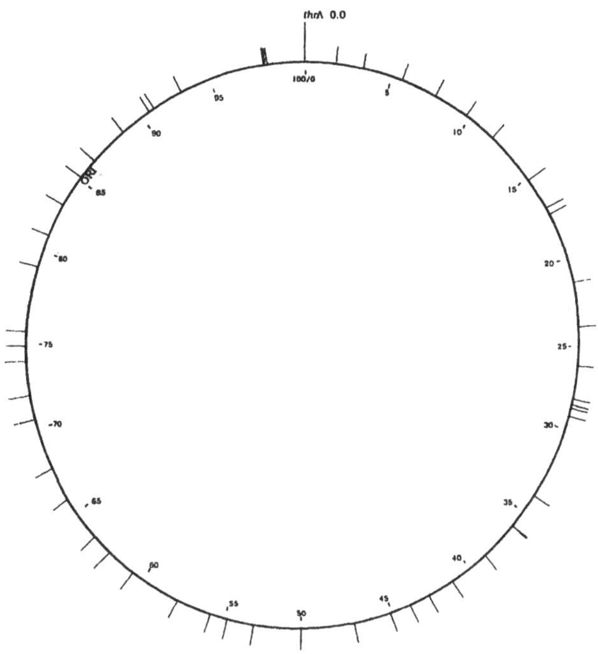

Figure 3-1. Preliminary *E. coli* genetic map.

TRANSDUCTION

As more and more genetic markers were placed on this map, it became obvious that the relative locations of nearby markers might not be precise, since interrupted mating could only be accurate to a distance corresponding to a few minutes. Another method of transferring genetic material from a donor to a recipient *E. coli* by a bacterial virus or phage was discovered, known as transduction (Lederberg *et al.*, 1951; Zinder and Lederberg, 1952). In the case of *E. coli* (Lennox, 1955), a mutant of the wild-type P1-phage was developed to carry a fixed amount of genetic material from any location of the donor chromosome.

This process consists of packaging a piece of *E. coli* DNA into the phage head accidentally. Thus, it occurs only rarely about one out of a thousand times. This hybrid phage is known as the transducing particle. If it infects the recipient *E. coli*, the small piece of DNA from the donor can thus be introduced into the recipient. The size of the donor DNA has a length of about two minutes, or roughly 1×10^5 base-pairs (bp), i.e. the original size of the P1-phage DNA. This length is ideal to complement the above-mentioned experimental location of genetic markers by conjugation.

A typical experiment involved the infection of the donor strain having the good gene, e.g. *thr*$^+$, with the P1-phage. The progeny phages together with the transducing particles were separated from the bacteria by centrifugation, and used to infect the recipient strain carrying *thr*$^-$, for example. Similar to the conjugation experiment, the streptomycin marker could be used to eliminate any contaminating donor strain, although not absolutely required. Transductants were then isolated on selecting plates and counted. Due to the much small size of the donor DNA fragment, the frequency of obtaining transductants would be much smaller than that of recombinants from conjugation experiments. Fortunately, it was usually much larger than the reversion rate.

If the distance between two genetic markers is less than two minutes, they can be transduced into the recipient simultaneously by one transducing particle. Let the markers in the donor be denoted by A+ and B+, and the corresponding ones in the recipient by A– and B–. In one experiment, let the experimentally measure number of transductants with A+ marker be [A+], and that of transductants with both A+ and B+ markers be [A+B+]. Then the co-transduction frequency is defined as:

$$\text{c.f.} = [A{+}B{+}] \,/\, [A{+}] \,.$$

Since transduction is a rare event, if the separation of the genetic markers A and B is larger than two minutes, c.f. = 0. Also, if genetic markers A and B are the same, c.f. = 1. The question is how to convert other values of c.f. into the physical distance between markers A and B. One simple-minded suggestion is to draw a straight line between the above two points (Fig. 3-2). However, there is no biological reasoning for this approach. Furthermore, its validity can be tested by considering three consecutive genetic markers, A, B and C. Experimentally, one can measure the co-transduction frequencies for A and B, B and C and A and C. The distances deduced from the straight line should be additive, i.e.

$$c.f._{AB} \longrightarrow D_{AB} ,$$

$$c.f._{BC} \longrightarrow D_{BC} ,$$

$$c.f._{AC} \longrightarrow D_{AC} ,$$

then, $$D_{AB} + D_{BC} = D_{AC} .$$

Consider an actual experiment with the following results (Singer *et al.*, 1965):

$$c.f._{AB} = 0.42 ,$$

$$c.f._{BC} = 0.57 ,$$

and $$c.f._{AC} = 0.19 .$$

Then, based on the straight line, we get.

$$D_{AB} = 1.16 \text{ min} ,$$

$$D_{BC} = 0.86 \text{ min} ,$$

and $$D_{AC} = 1.62 \text{ min} .$$

However, $$D_{AB} + D_{BC} = 2.02 \text{ min} \neq 1.62 \text{ min} = D_{AC} .$$

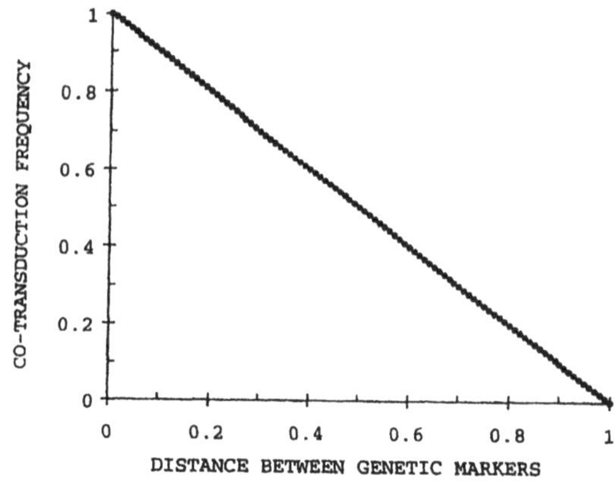

Figure 3-2. Straight line relation of co-transduction frequency and distance between genetic markers.

For this reason, many transduction experiments involving three nearby genetic markers have been performed, known as three-point analyses. It is thus essential to derive the biologically relevant curve connecting the two above-mentioned end points. Furthermore, the accumulated data on many three-point analysis experimental measurements can be used to test the usefulness of the theoretical curve.

MODEL FOR RANDOM GENERALIZED TRANSDUCTION

The accidental packaging of a piece of *E. coli* chromosomal DNA into the P1-phage head implies that not only the piece has a fixed length of two minutes but also that the piece is cut out randomly over the entire chromosome. The latter property is commonly referred as generalized. Let the recipient *E. coli* carry a defective gene designated by A–, and the donor *E. coli* the good gene A+. For a typical transduction experiment, we would like to calculate the chance of obtaining a transductant carrying the A+ gene. Since the *E. coli* chromosome is circular, two or any even number of cross-overs are required to maintain the viability of the transductant. Since cross-overs are rare events, we only need to consider two cross-overs as shown in Fig. 3-3.

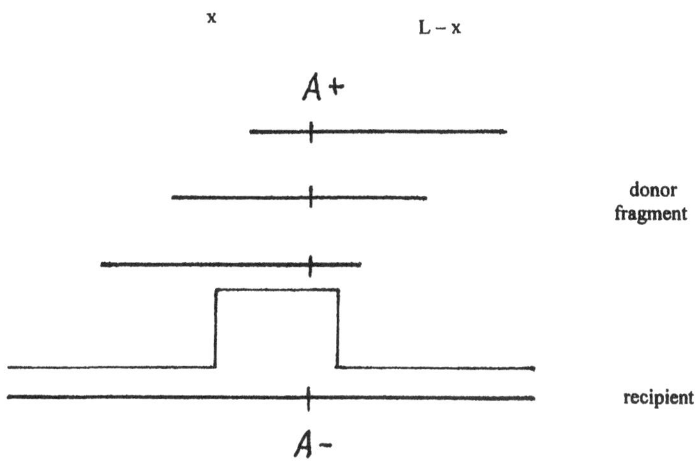

Figure 3-3. The basic mechanism of obtaining an A+ transductant.

Cross-overs arc proportional to the length of the involved DNA segment, and the constant of proportionality can be denoted by μ. Since we have no idea of the distance from the left end of the donor DNA fragment to the marker A+, we shall designate it as x. However, we know that the entire length of the donor DNA fragment is fixed and can be denoted by L. Thus, the distance from the marker A+ to the right end of the donor DNA fragment is L-x. In random generalized transduction, a uniform distribution of various pieces of donor DNA fragments is assumed and the distribution constant is denoted by c. Then, the chance of obtaining an A+ transductant can be calculated by integrating over all relevant values of x, i.e. from 0 to L:

$$[A+] = \int_0^L c\,\mu x\,\mu(L - x)\,dx$$

$$= c\,\mu^2 (Lx^2/2 - x^3/3)\Big|_0^L$$

$$= c\,\mu^2 (L^3/2 - L^3/3)$$

$$= c\,\mu^3 L^3/6\,.$$

The values of c and μ will depend on the conditions of the experiment and can vary from experiment to experiment. Therefore, it is difficult to predict the value of [A+]. Fortunately, we are not interested in the value of [A+] but rather the ratio of [A+B+] / [A+] or the co-transduction frequency. The basic mechanism of obtaining A+B+ transductants is illustrated in Fig. 3-4. We need one cross-over to the left of A and another to the right of B.

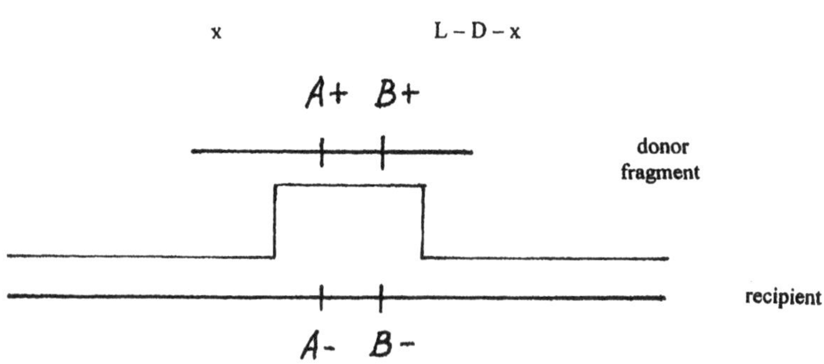

Figure 3-4. Basic mechanism of obtaining an A+B+ transductant.

Let the distance between genetic markers A and B be denoted by D. Then a similar calculation as before gives:

$$\{A+B+\} = \int_0^{L-D} c\,\mu x\,\mu(L-D-x)\,dx$$

$$= c\,\mu^2\,(L-D)^3/6\,.$$

Therefore, the co-transduction frequency is simply:

$$\text{c.f.} = [A+B+]/[A+] = (1-D/L)^3\,.$$

which is independent of c or μ (Wu, 1966). As long as the measurement of [A+] and [A+B+] are carried out in the same experiment, there is no need to determine c nor μ. This result is plotted in Fig. 3-5, quite different from the straight line shown in Fig. 3-2.

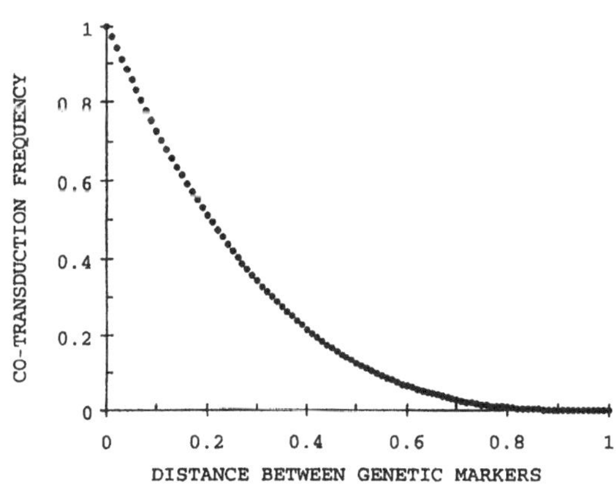

Figure 3-5. Biological relevant relation of co-transduction frequency and distance between genetic markers in generalized random transduction.

For a given experimentally measured co-transduction frequency, the straight line in Fig. 3-2 and the curve in Fig. 3-5 give very different distances between the two genetic markers. In the case of the experiment mentioned before, we have:

$$D_{AB} = 0.50 \text{ min} ,$$

$$D_{BC} = 0.34 \text{ min} ,$$

and

$$D_{AC} = 0.85 \text{ min} .$$

Therefore, the sum of D_{AB} and D_{BC} is 0.84 min close to 0.85 min, D_{AC}.

The estimated distances between genetic markers using the biological relevant curve (Fig. 3-5) are much smaller than those estimated by the straight line relation (Fig. 3-2). With this mathematical modeling, three-point analysis can give consistent and accurate distances between the three consecutive genetic markers as discussed below.

THREE-POINT ANALYSIS

Transduction experiments were performed for numerous genetic markers of *E. coli*, some involve three nearby consecutive ones. These experiments have been referred to as three-point analysis. Let us assume that A, B and C are the three nutrient supplements required for the growth for the recipient strain which is thus denoted as A–B–C–. On the other hand, the donor strain does not require these for growth, so that it is designated as A+B+C+. In a typical experiment, P1-phage is first grown with the donor strain and then harvested. After mixing with the recipient strain, they are then plated out on plates with supplements B and C, C and A, or A and B, to select for A+, B+ and C+ transductants respectively. Streptomycin can be used to eliminate the background growth of the donor strain as in a conjugation experiment.

After a day or two, transductants will form colonies on these three plates. Each colony is then picked by sterile tooth-picks and patched onto plates with the same supplements in a pattern shown in Fig. 3-6. Usually, 100 patches are made, with 50 patches per plate.

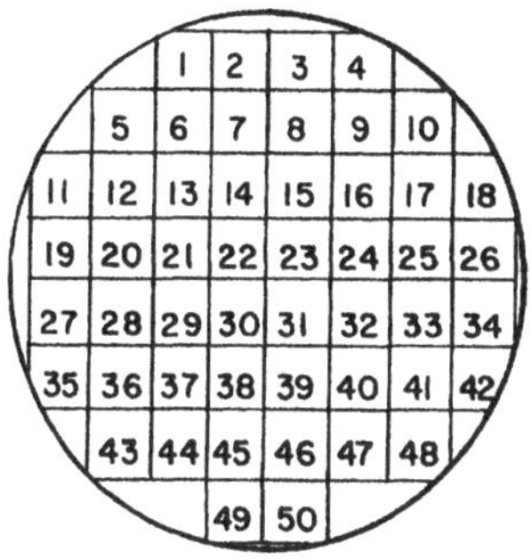

Figure 3-6. Patch pattern for 50 patches per plate.

These patches grow out much faster, usually over-night. Each plate is then replica plated onto other plates, using a velvet stamp slightly smaller than the surface of the plate. For the original plate with supplements B and C, four replica plates are used, one with supplement B only, one with C only, one with no supplement, and the last plate with both supplements. These plates will score for transductants with the following genetic markers:

plate with supplement B only: A+B+C+, A+B–C+;

plate with supplement C only: A+B+C+, A+B+C–;

plate with no supplement: A+B+C+;

and plate with both supplements: A+B+C+, A+B+C–, A+B–C+, A+B–C–.

Similarly, for the original plate with supplements C and A, also four replica plates are used:

plate with supplement C only: A+B+C+, A+B+C–;

plate with supplement A only: A+B+C+, A–B+C+;

plate with no supplement: A+B+C+;

and plate with both supplements: A+B+C+, A+B+C–, A–B+C+, A–B+C–.

For the original plate with supplements A and B, the four replica plates are:

plate with supplement A only: A+B+C+, A–B+C+;

plate with supplement B only: A+B+C+, A+B–C+;

plate with no supplement: A+B+C+;

and plate with both supplements: A+B+C+, A+B–C+, A–B+C+, A–B–C+.

After sorting out the above results, twelve measurements are available:

[A+]: [A+B–C–]

[A+B+C–]

[A+B–C+]

[A+B+C+]

[B+]: [A–B+C–]

[A+B+C–]

[A–B+C+]

[A+B+C+]

[C+]: [A–B–C+]

[A–B+C+]

[A+B–C+]

[A+B+C+]

Just as before, ratios are usually reported. In these studies, however, there are only two unknowns to be determined, namely, the distance between markers A and B, and that between B and C. We shall denote them as S and T respectively. As a result, three-point analysis can provide very accurate estimations of distances between genetic markers.

The equations for these ratios can be derived as follows. In order to obtain transductants with [A+B–C+], we need a cross-over to the left of A, one between A and B, another between B and C, and a fourth to the right of C. Therefore, the experimentally measured number will be proportional to μ^4. Dividing by [A+] or [C+] both being proportional to μ^2, we get a ratio proportional to μ^2. Since μ is a small number, μ^2 is negligible, usually much less than 0.01. This finding has been used to order the three genetic markers. Thus, among the twelve experimentally measured ratios, two should be zero or less than 0.01. The other ten will then be functions of two non-dimensional ratios:

$$\alpha = S / L ,$$

and $$\beta = T / L .$$

To derive these equations, refer to Figs. 3-3 and 3-4. First, we have:

$$[A+] = [B+] = [C+] ,$$

since they are all equivalent to each other. If we have three genetic markers A, B and C in that order, instead of two genetic markers A and B shown in Fig. 3-4, to get A+B+C+ transductants, we need a cross-over to the left of A and another to the right of C. Thus, we have:

$$[A+B+C+] / [A+] = (1 - S / L - T / L)^3$$

$$= (1 - \alpha - \beta)^3 .$$

If we consider A+B+ transductants, the third genetic marker can be C+ or C–. Therefore, by definition, we have:

$$[A+B+] / [A+] = [A+B+C+] / [A+] + [A+B+C-] / [A+] ,$$

or $$(1 - \alpha)^3 = (1 - \alpha - \beta)^3 + [A+B+C-] / [A+] .$$

or $$[A+B+C-] / [A+] = (1 - \alpha)^3 - (1 - \alpha - \beta)^3 .$$

Then among [A+], the sum of the four ratios must equal to one, with one of the four being zero. Thus, we have:

$$[A{+}B{-}C{-}]\,/\,[A{+}] = 1 - (1-\alpha)^3,$$

$$[A{+}B{+}C{-}]\,/\,[A{+}] = (1-\alpha)^3 - (1-\alpha-\beta)^3,$$

$$[A{+}B{-}C{+}]\,/\,[A{+}] = 0,$$

$$[A{+}B{+}C{+}]\,/\,[A{+}] = (1-\alpha-\beta)^3,$$

$$[A{-}B{+}C{-}]\,/\,[B{+}] = 1 + (1-\alpha-\beta)^3 - (1-\alpha)^3 - (1-\beta)^3,$$

$$[A{+}B{+}C{-}]\,/\,[B{+}] = (1-\alpha)^3 - (1-\alpha-\beta)^3,$$

$$[A{-}B{+}C{+}]\,/\,[B{+}] = (1-\beta)^3 - (1-\alpha-\beta)^3,$$

$$[A{+}B{+}C{+}]\,/\,[B{+}] = (1-\alpha-\beta)^3,$$

$$[A{-}B{-}C{+}]\,/\,[C{+}] = 1 - (1-\beta)^3,$$

$$[A{-}B{+}C{+}]\,/\,[C{+}] = (1-\beta)^3 - (1-\alpha-\beta)^3,$$

$$[A{+}B{-}C{+}]\,/\,[C{+}] = 0,$$

and $$[A{+}B{+}C{+}]\,/\,[C{+}] = (1-\alpha-\beta)^3.$$

The ten non-zero experimentally measurable ratios are all functions of two variables, α and β. Therefore, again by least square fitting of the experimental data to the theoretical equations, we should be able to determine α and β very accurately. On multiplying by L, the length of the

DNA carried by the transducing particle of two minutes or 9.28 X 10^4 bp, α and β can give the distances between genetic markers A and B and that between B and C.

AN EXAMPLE

Numerous three-point analysis experimental data are available in the literature. A typical example is illustrated here. The three sets of four replica plates of patches are shown in Figs. 3-7 to 3-9. These numbers of patches are then converted to ratios given in Table 3-1 (Singer *et al.*, 1965).

Table 3-1. Ratios from a three-point analysis experiment.

	[A+B–C–]	[A+B+C–]	[A+B–C+]	[A+B+C+]	[A–B+C–]	[A–B+C+]	[A–B–C+]
{A+]	0.51	0.30	0.00	0.19			
[B+]		0.20		0.17	0.20	0.43	
[C+]			0.00	0.14		0.43	0.43

It is obvious from Table 3-1 that experimental errors are not avoidable. For example, theoretically

$$[A+B+C-] / [A+] = [A+B+C-] / [B+],$$

but the experimental measured ratios are 0.30 and 0.20 respectively. It is for this reason that least square fitting to the ten non-zero experimental data points can give very accurate values of α and β. Initial estimations of α and β can be obtained from [A+B–C–] / [A+] and [A–B–C+] / [C+], i.e.

$$0.51 = 1 - (1 - \alpha)^3,$$

and

$$0.43 = 1 - (1 - \beta)^3.$$

Therefore, $\alpha = 0.21$ and $\beta = 0.17$.

plate with supplement B only: A+B+C+, A+B–C+

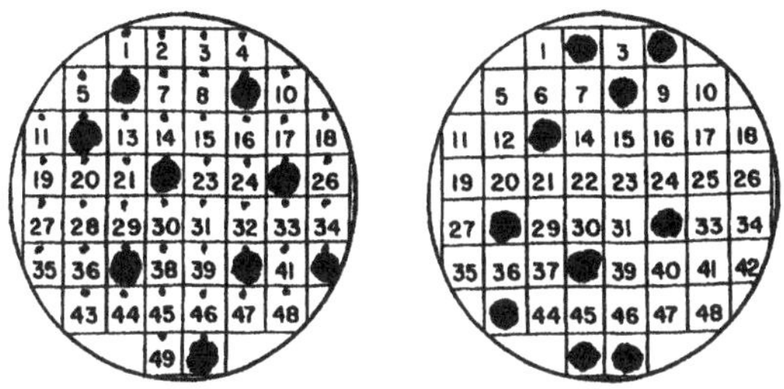

plate with supplement C only: A+B+C+, A+B+C–

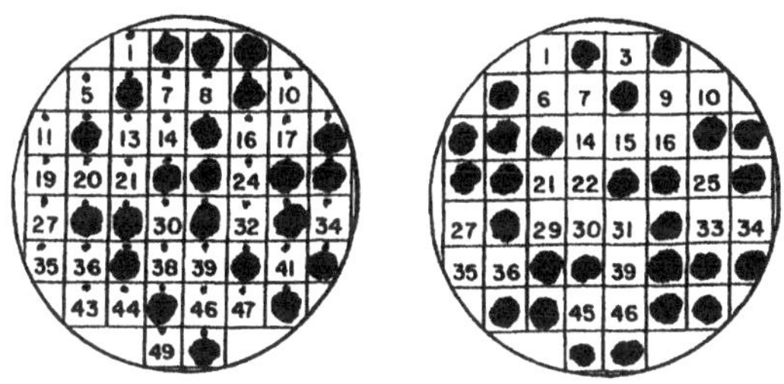

plate with no supplement: A+B+C+

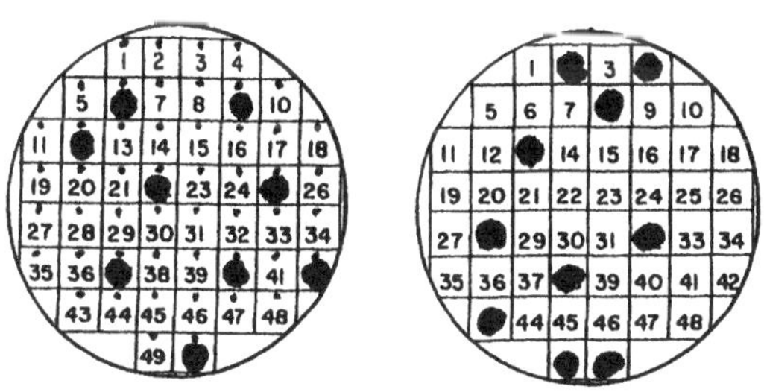

Figure 3-7. Replica plating for colonies selected with supplements B and C.

plate with supplement C only: A+B+C+, A+B+C–

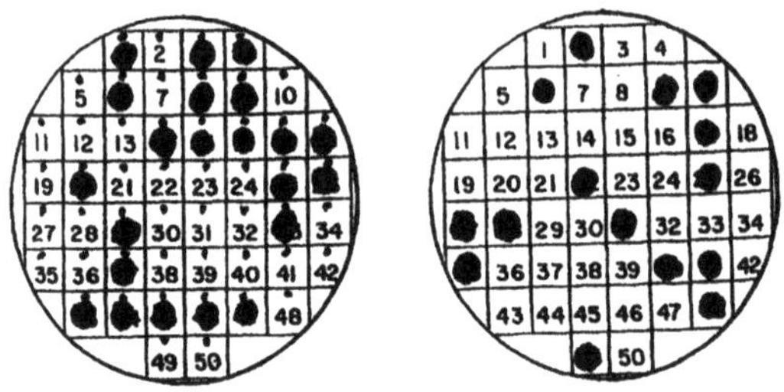

plate with supplement A only: A+B+C+, A–B+C+

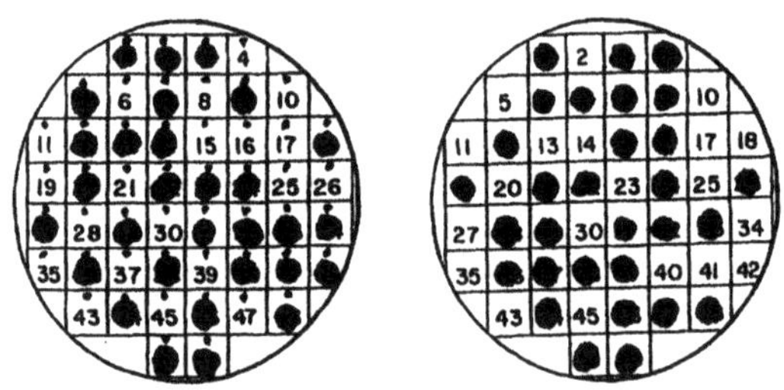

plate with no supplement: A+B+C+

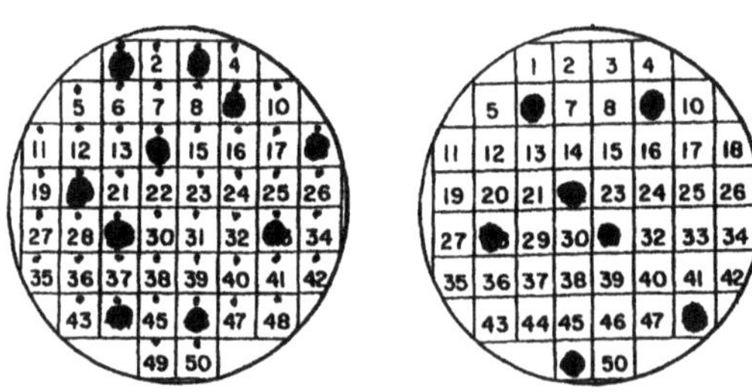

Figure 3-8. Replica plating for colonies selected with supplements C and A.

plate with supplement A only: A+B+C+, A–B+C+

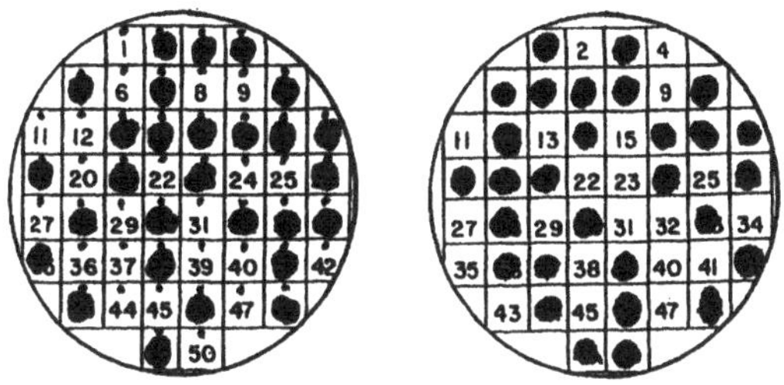

plate with supplement B only: A+B+C+, A+B–C+

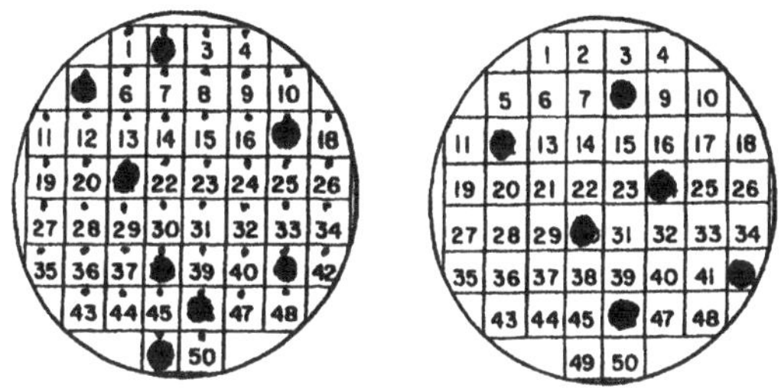

plate with no supplement: A+B+C+

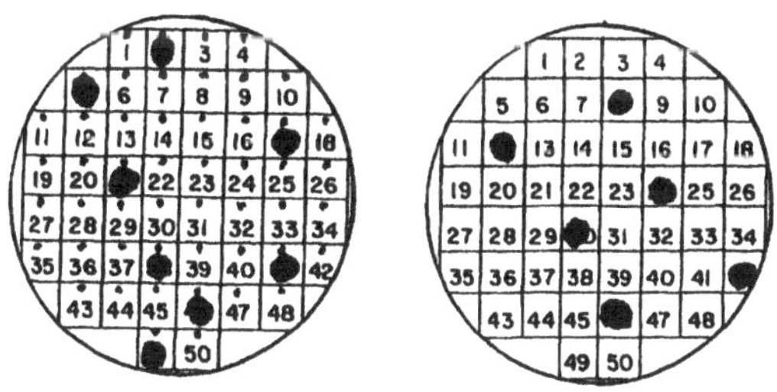

Figure 3-9. Replica plating for colonies selected with supplements A and B.

We can start with these initial values of α and β to obtain more accurate values by least square fitting (discussed in Chapter 2) to the ten non-zero values in Table 3-1. The sum of squares of the differences between theory and experiment, in this case, can be denoted as $S(\alpha,\beta)$. First, β is fixed at 0.17, and α is varied to calculate the value of S, as shown in Table 3-2.

Table 3-2. Adjusting α to find the minimum value of $S(\alpha,\beta)$ for $\beta = 0.17$.

α	$S(\alpha,\beta)$
0.250	0.020107
0.253	0.019916
0.254	0.019890
0.255	0.019882
0.256	0.019893
0.257	0.019921
0.260	0.020118

Therefore, we can use $\alpha = 0.255$, and then vary β to get a minimum value of $S(\alpha,\beta)$ as shown in Table 3-3.

Table 3-43. Adjusting β to find the minimum value of $S(\alpha,\beta)$ for $\alpha = 0.255$.

β	$S(\alpha,\beta)$
0.170	0.019882
0.171	0.019861
0.172	0.019865
0.173	0.019894
0.174	0.019949

Finally, by least square fitting, we get:

$$\alpha = 0.255 ,$$

and $$\beta = 0.171 .$$

Multiplying the above values of α and β by the length of the DNA in the transducing particle or 2 min, we have the separation between genetic markers A and B being approximately 0.51 min and that between markers B and C 0.34 min.

Escherichia coli GENETIC MAP

Taylor (1970) started to examine the genetic markers of *E. coli* carefully and constructed a linkage map. At that time, the entire length of the *E. coli* chromosome was assumed to be 90 minutes in length. That map was re-calibrated in 1976 by Bachmann *et al.* As more and more genetic markers were identified, Bachmann (1983, 1990) have been incorporating all available genetic markers of *E. coli* into the linkage map using tranduction experiments and the above derived equations. She had positioned a total of 1,403 of them on the *E. coli* chromosome by 1990. During the past ten years, even more genetic markers have been identified (see the special website at Yale University, <http://cgsc.biology.yale.edu >). As an example, the section of the *E. coli* chromosomal map from 0.0 minute to 1.0 minute is shown in Fig. 3-10. This information has provided a useful guide to the sequencing of *E. coli* genome (see, for example, Itoh *et al.*, 1996). Recently, both the traditional map (Berlyn, 1998) and the physical map (Rudd, 1998) have been summarized. Other websites can be found from the above website and are also listed by Rudd (2000).

The physical map (Rudd, 1998) consists of restriction enzyme sites and positions of cloned segments of *E. coli* chromosome in lambda phage EMBL4 (Kohara *et al.*, 1987). It has been correlated with the 4,639,221 bp DNA sequence of *E. coli* K-12 strain MG1655 version M52 (Blattner *et al.*, 1997). The *thr A* gene has been sequenced (Katinta *et al.*, 1980) and was physically positioned near the beginning of the DNA sequence by convention (Blattner *et al.*, 1997)

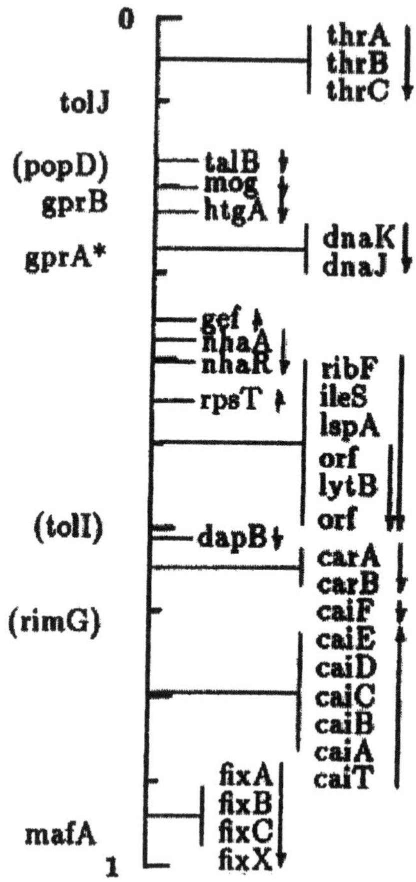

Figure 3-10. Traditional map of 0.0 minute to 1.0 minute of the *E. coli* chromosome (from <http://mmbr.asm.org/content/vol62/issue3/images/large/mr0380033lp1.jpeg>).

FUTURE WORK

As pointed out by Rudd (2000), *E. coli* has been an old workhorse in molecular biology, as well as in biochemistry, cell biology, genetics, etc. With the complete knowledge of its genome, more intriguing questions can be asked. There are many open reading frames (ORF's) which can code for proteins with unknown functions. These proteins can be synthesized *in vitro* and their functions investigated. The development of a new technology known as DNA chips can thus help our understanding of numerous biological processes carried out by *E. coli* (Selinger *et al.*, 2000). In their study, they used a 544 X 544 grid of 24 X 24 μm regions to simultaneously measure RNA abundance for large numbers of genes under various growth conditions as a function of time. The information contents of these chips are extensive. How to analyze such information remains a challenge to both biologists and mathematicians.

REFERENCES

Bachman BJ (1983) Linkage map of *Escherichia coli* K-12, edition 7. *Microbiol. Rev.*, **47**, 180-230.

Bachmann BJ (1990) Linkage map of *Escherichia coli* K-12, edition 8. *Microbiol. Rev.*, **54**, 130-197.

Bachman BJ, Low KB and Taylor AL (1976) Recalibrated linkage map of *Escherichia coli* K-12. *Bacteriol. Rev.*, **40**, 116-167.

Berlyn MKB (1998) Linkage map of *Escherichia coli* K-12, edition 10: the traditional map. *Microbiol. Mol. Biol. Rev.*, **62**, 814-984.

Blattner FR, Plunkett G, Bloch CA, Perna NT, Burland V, Riley M, Collado-Vides J, Glasner JD, Rode CK, Mayhew GF, Gregor J, Davis NW, Kirkpatrick HA, Goeden MA, Rose DJ, Mau B and Shao Y (1997) The complete genome sequence of *Escherichia coli* K-12. *Science,* **277**, 1453-1462.

Itoh T, Aiba H, Baba T, Hayashi K, Inada T, Isono K, Kasai H, Kimura S, Kitakawa M, Kitakawa M, Makino K, Miki T, Mizobuchi K, Mori H, Mori T, Motomura K, Nakade S, Nakamura Y, Nashimoto H, Nishio Y, Oshima T, Sato N, Sampei G, Seki Y, Sivasunddaram S, Tagami H, Takeda J, Takemoto K, Wada C, Yamamoto Y and Horiuchi T (1996) A 460-kb DNA sequence of *Escherichia coli* K-12 genome corresponding to the 40.1-50.0 min region of the linkage map. *DNA Res.*, **3**, 379-392.

Katinka M, Cossart P, Sibilli L, Saint-Girons I, Chalvignac MA, Le Bras G, Cohen GN and Yaniv M (1980) Nucleotide sequence of the *thr A* gene of *Escherichia coli. Proc. Nat. Acad. Sci. USA,* **77**, 5730-5733.

Kohara Y, Akiyama K and Isono K (1987) The physical map of the whole E. coli chromosome: application of a new strategy for rapid analysis and sorting of large genomic library. *Cell,* **50,** 495-508.

Lederberg J, Lederberg EM, Zinder ND and Lively ER (1951) Recombination analysis of bacterial heredity. *Cold Spring Harbor Symp. Quant. Biol.,* **16,** 413-441.

Lederberg J and Tatum EL (1946) Gene recombination in *Escherichia coli. Nature,* **138,** 558.

Lennox E (1955) Transduction of linked genetic characters of the host by bacteriophage P1. *Virology,* **1,** 190-206.

Rudd KE (1998) Linkage map of *Escherichia coli* K-12, edition 10: the physical map. *Microbiol. Mol. Biol. Rev.,* **62,** 985-1019.

Rudd KE (2000) New tools for an old workhorse. *Nature Biotech.,* **18,** 1241-1242.

Selinger DW, Cheung KJ, Mei R, Johansson EM, Richmond CS, Blattner FR, Lockhart DJ and Church GM (2000) RNA expression analysis using a 30 base pair resolution *Escherichia coli* genome array. *Nature Biotech.,* **18,** 1262-1268.

Singer ER, Beckwith JR and Brenner S (1965) Mapping of suppressor loci in *Escherichia coli. J. Mol. Biol.,* **14,** 153-166.

Taylor AL (1970) Current linkage map of *Escherichia coli.* ***Bacteriol. Rev.,*** **34,** 155-175.

Wu TT (1966) A model for three-point analysis of random general transduction. *Genetics,* **54,** 405-410.

Zinder ND and Lederberg J (1952) Genetic exchange in *Salmonella. J. Bacteriol.,* **64,** 679-699.

EXERCISE

Calculate the distances between genetic markers D and E and between E and F based on the following experimental result of another three-point analysis:

	[D+E–F–]	[D+E+F–]	[D+E–F+]	[D+E+F+]	[D–E+F–]	[D–E+F+]	[D–E–F+]
[D+]	0.75	0.19	0.00	0.06			
[E+]		0.18		0.10	0.31	0.41	
[F+]			0.00	0.07		0.52	0.41

Obtain their initial values, and then try to improve their accuracy by least square fitting.

CHAPTER 4

ENZYME KINETICS

INTRODUCTION

Biochemists have isolated many enzymes from various biological sources (Dixon and Webb, 1964), and molecular biologists have analyzed their amino acid and nucleotide sequences together with their three-dimensional structures, in order to understand their functions. Enzymes catalyze certain chemical reactions, and have evolved over millions of years to carry out their functions under required conditions very efficiently. The properties of enzymes are thus specified by the rates of reactions they mediate. At the same time, they may also promote reactions of similar chemicals but to a less degree of efficiency.

For our present discussion, one of the glycerol dehydrogenases will be considered (Fig. 4-1). It removes two hydrogen atoms from glycerol to form dihydroxyacetone. The hydrogen atoms are delivered to a co-enzyme, nicotine adenine dinucleotide (NAD). There are two nearby pockets on the enzyme, one for glycerol and the other for NAD. The two small molecules can bind to the enzyme sequentially. Then, they react and the products, dihydroxyacctone and reduced NAD, will dissociate from the enzyme molecule.

It is not easy to measure the concentration of dihydroxyacetone. Thus, the rate of the enzymatic reaction is usually measured by the amount of reduced NAD as a function of time. This molecule has a characteristic absorption of light with wave-length at 340 nm. In a typical experiment, various amounts of glycerol dehydrogenase, nicotine adenine dinucleotide and glycerol are mixed in a buffer promoting this reaction at a pH not necessarily physiological. The solution is then loaded into a UV-transparent glass cuvette scanned by a spectrophotometer at 340 nm and recorded on a time chart. The slope of the curve gives the velocity of the enzymatic reaction for the given concentrations of the enzyme, co-enzyme and the substrate.

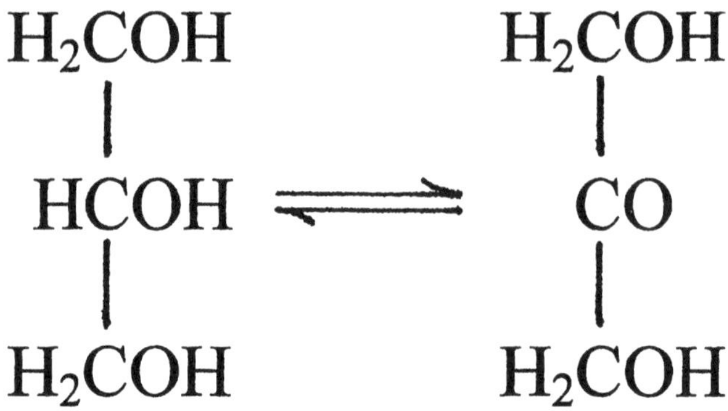

Figure 4-1. Conversion of glycerol to dihydroxyacetone by glycerol dehydrogenase.

SIMPLIFIED REACTION

If we represent glycerol dehydrogenase by E, NAD by C, glycerol by S, dihydroxyacetone by P and reduced NAD by C̲, the above mentioned reaction can be summarized as in Fig. 4-2.

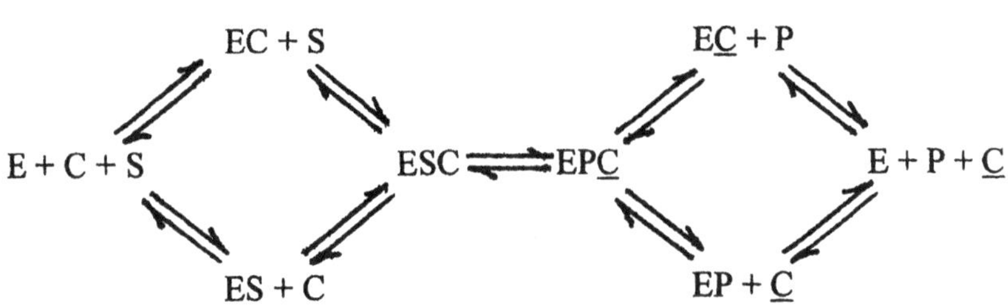

Figure 4-2. Schematic representation of an enzymatic reaction.

There are 18 kinetic constants (represented by the 18 arrows in Fig. 4-2) to be measured. Since this is obviously not possible for most enzymes, drastic simplification is necessary. For most enzymatic reactions, the co-enzyme NAD is omitted even though the concentration of reduced NAD is actually measured. Thus, sufficient amount of NAD must be added to saturate the enzyme with the co-enzyme. Then the production of the product is considered to be irreversible (Fig. 4-3).

$$E + S \underset{k_{-1}}{\overset{k_1}{\rightleftharpoons}} ES \overset{k_2}{\longrightarrow} E + P$$

Figure 4-3. Simplified enzymatic reaction.

As a result, 18 kinetic constants are reduced to three. This simplified reaction is usually referred to as the Michaelis-Menten kinetics (Michaelis and Menten, 1913). Rate equations are needed to determine them.

RATE EQUATIONS

From Fig. 4-2, the rate equations can be written down using the directions of the arrow:

$$d[E] / dt = -k_1 [E][S] + k_{-1} [ES] + k_2 [ES] ,$$

$$d[S] / dt = -k_1 [E][S] + k_{-1} [ES] ,$$

$$d[ES] / dt = k_1 [E][S] - k_{-1} [ES] - k_2 [ES] ,$$

and

$$d[P] / dt = k_2 [ES] .$$

They satisfy the initial conditions at $t = 0$, $[E] = [E]_0$, $[S] = [S]_0$, $[ES] = 0$ and $[P] = 0$. This set of ordinary differential equations is non-linear, since three of the equations contain terms with the product of two unknowns. However, we have:

$$d[E] / dt + d[ES] / dt = 0 ,$$

or

$$[E] + [ES] = [E]_0 .$$

This simply states that enzyme molecules are conserved.

SIMPLE ENZYME KINETICS

In simple enzyme kinetics, it is assumed that the formation of the enzyme-substrate complex, ES, occurs very rapidly. Therefore, it is assumed that

$$d[ES] / dt = 0 .$$

so that the differential equation involving d[ES]/dt is reduced to an algebraic equation:

$$k_1 [E] [S] - k_{-1} [ES] - k_2 [ES] = 0 .$$

The second assumption in simple enzyme kinetics is that the concentration of the substrate, [S], hardly changes, i.e.:

$$[S] = [S]_0 ,$$

at all times. This is a reasonable assumption if:

$$[S]_0 >> [E]_0 .$$

In most *in vitro* experimental measurements, $[S]_0$ is of the order of 10^{-6} M or larger while $[E]_0$ is of the order of 10^{-8} M. However, in metabolic pathways involving a number of consecutive enzymes, this assumption may not be valid. With this assumption, the rate equation involving d[S]/dt must be ignored.

As a result, we have the following set of algebraic equations:

$$[E] + [ES] = [E]_0 ,$$

$$k_1 [E] [S]_0 - k_{-1} [ES] - k_2 [ES] = 0 ,$$

and

$$V = k_2 [ES] ,$$

where V is the velocity of the reaction and is equal to d[P]/dt.. We can thus express [E] in terms of $[E]_0$ and [ES], and express [ES] in terms of $[E]_0$ and $[S]_0$, i.e.

$$[E] = [E]_0 - [ES] ,$$

which can be substituted into the equation with the kinetic constants to give:

$$k_1 ([E]_0 - [ES]) [S]_0 - (k_{-1} + k_2) [ES] = 0 .$$

From this equation, [ES] can be expressed in terms of the three kinetic constants and the initial concentrations of the enzyme and substrate:

$$[ES] = [E]_0 [S]_0 / \{[S]_0 + (k_{-1} + k_2) / k_1\} .$$

Finally, we can calculate the velocity of the simple enzyme kinetic reaction:

$$V = V_{max} [S]_0 / \{[S]_0 + K_m\} ,$$

with $$V_{max} = k_2 [E]_0 ,$$

and $$K_m = (k_{-1} + k_2)/k_1 .$$

This equation is essentially identical to that for equilibrium saturation of myoglobin with oxygen (see Chapter 2).

As an example, one of the glycerol dehydrogenases (McGregor *et al.*, 1974) has been purified, and its enzymatic activity measured in the absence of any cation (Table 4-1). For a concentration of the co-enzyme, NAD, at 0.343 mM, the velocity of the reaction, V, was expressed in mM of NADH formed per minute per mg of enzyme, and the initial concentration of glycerol, $[S]_0$, in mM.

Table 4-1. Experimentally determined V at various $[S]_0$ for a glycerol dehydrogenase as measured from one of the graphs published by McGregor *et al.* (1974).

$[S]_0$ (mM)	V (mM NADH min^{-1} mg^{-1})
51.9	19.27 - 20.04
70.6	22.99 - 24.27
98.7	28.25 - 29.76
182.5	37.59 - 39.06

Similarly, three different plots are commonly used by biochemists to determine V_{max} and K_m from such experimental measurements. Fig. 4-4 shows the plot of V against $[S]_0$ as a hyperbola. Fig. 4-5 shows that of 1/V against $1/[S]_0$, and Fig. 4-6 that of V against $V/[S]_0$, both as straight lines. Sometimes, the axes may be exchanged.

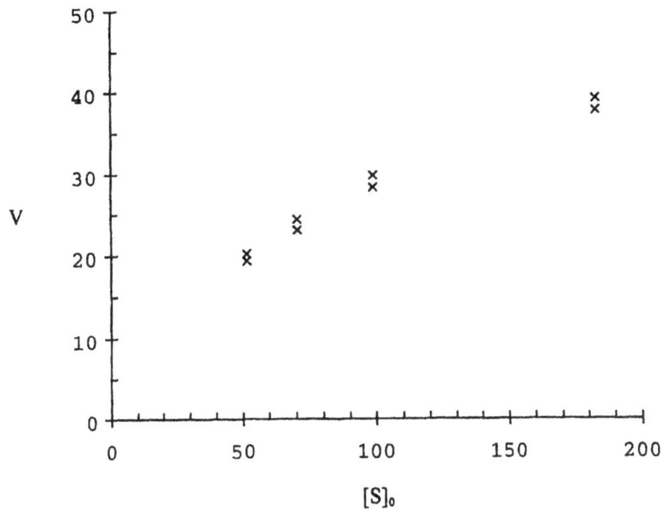

Figure 4-4. V plotted against $[S]_0$.

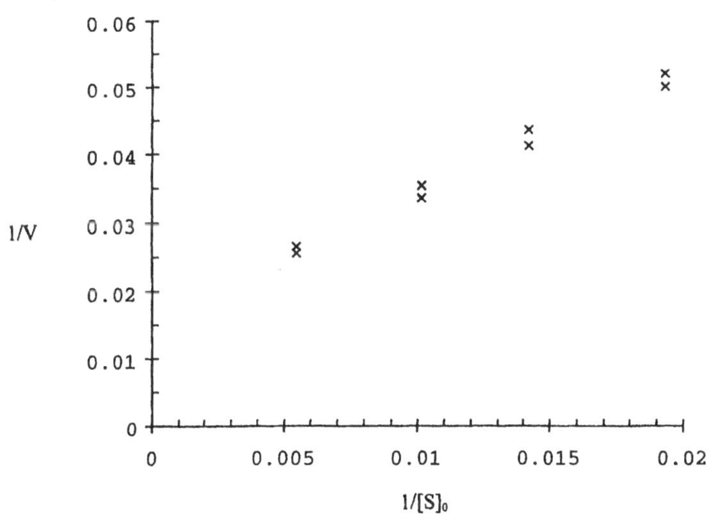

Figure 4-5. 1/V plotted against $1/[S]_0$.

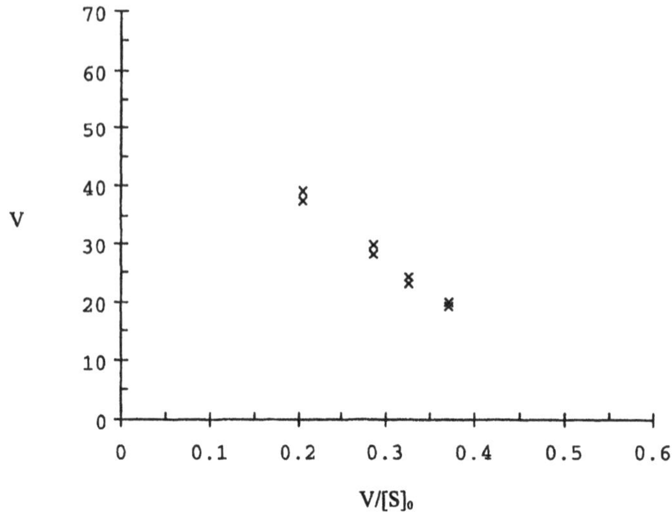

Figure 4-6. V plotted against $V/[S]_0$.

One immediately realizes that not all the original three kinetic constants, k_1 , k_{-1} and k_2 can be determined in simple enzyme kinetics. If the initial enzyme concentration, $[E]_0$, is known, then V_{max} can be used to determine k_2 , since:

$$k_2 = V_{max} / [E]_0 .$$

However, K_m is a combination of all three constants, and we cannot solve for k_1 and k_{-1} separately. This is the consequence of the two simplifying assumptions reducing the original four ordinary differential equations to three algebraic equations. In order to understand the problems caused by these assumptions, we need to examine the initial conditions carefully. At time zero, [ES] = 0. However, in the "steady state" of simple enzyme kinetics, [ES] has a non-zero value. Similarly, the enzyme concentration, [E], is also discontinuous for $t = 0-$ and $t = 0+$. As to the concentration of the product, [P], at $t = 0$, its slope should be zero, not at a finite value, since $d[P]/dt = k_2$ [ES] and [ES] = 0. These discontinuities of [S], [E], [ES] and [P] are illustrated in Fig. 4-7.

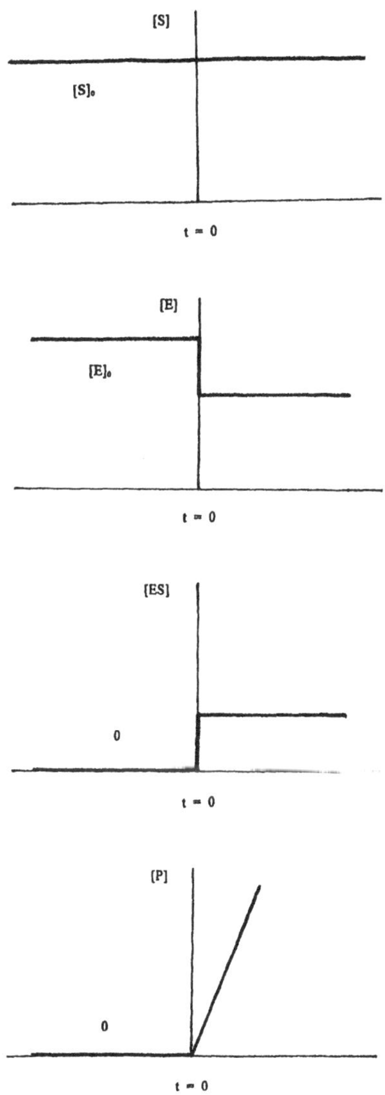

Figure 4-7. Discontinuities of [S], [E], [ES] and [P] just before and after t = 0.

Mathematically, it is essential to find solutions suitable of fitting into this short time interval. The assumption causing this problem is setting:

$$d[ES] / dt = 0 .$$

The other assumption of $[S] = [S]_0$ may still be acceptable as long as:

$$[S]_0 >> [E]_0 .$$

In this case, the first-order non-linear ordinary differential equation is reduced to a first-order linear ordinary different equation:

$$d[ES] / dt = k_1 [E] [S]_0 - (k_{-1} + k_2) [ES]$$

$$= k_1 [E]_0 [S]_0 - (k_1 [S]_0 + k_{-1} + k_2) [ES] .$$

TRANSITION REGION

To solve this equation, multiple it with the integrating factor:

$$\exp[(k_1 [S]_0 + k_{-1} + k_2) t]$$

so that terms contain [ES] can be combined. Therefore,

$$\frac{d\{[ES] \exp[(k_1[S]_0 + k_{-1} + k_2)t\}}{dt} = k_1 [E]_0 [S]_0 \exp[(k_1[S] + k_{-1} + k_2)t] .$$

On integrating with respect to t, we get:

$$[ES] \exp[(k_1[S]_0 + k_{-1} + k_2)t] = \frac{k_1 [E]_0 [S]_0}{k_1[S]_0 + k_{-1} + k_2} \exp[(k_1[S]_0 + k_{-1} + k_2)t]$$

$$+ \text{ constant} .$$

The constant can be determined with the initial condition, i.e. at $t = 0$, $[ES] = 0$. Thus, the constant equals to $-k_1 [E]_0 [S]_0 / (k_1[S]_0 + k_{-1} + k_2)$. As a result,

$$[ES] = \frac{k_1 [E]_0 [S]_0}{k_1[S]_0 + k_{-1} + k_2} \{1 - \exp-[(k_1[S]_0 + k_{-1} + k_2)t]\}$$

For large values of t, the exponential term approaches zero, and [ES] approaches the value found in simple enzyme kinetics.

Similarly, [E] also has this exponential transition from $[E]_0$ to the "steady state" value in simple enzyme kinetics, since $[E] + [ES] = [E]_0$.

To calculate [P], we need to integrate the rate equation:

$$\frac{d[P]}{dt} = k_2 [ES] = \frac{V_{max} [S]_0}{[S]_0 + K_m} \{1 - \exp-[k_1([S]_0 + K_m)t\} ,$$

and satisfy the initial condition of $[P] = 0$ at $t = 0$.

$$[P] = \frac{V_{max} [S]_0}{[S]_0 + K_m} \{t - \frac{1 - \exp-[k_1([S]_0 + K_m)t]}{k_1([S]_0 + K_m)}\} .$$

These transition regions are illustrated in Fig. 4-8 for small time.

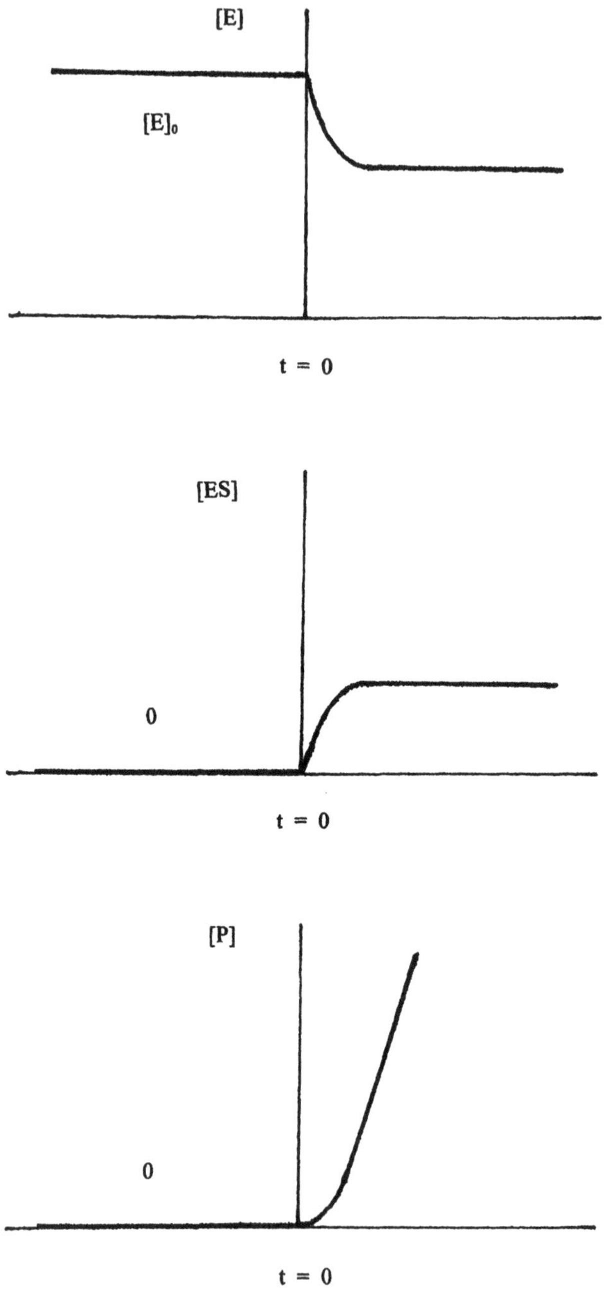

Figure 4-8. Transition regions for [E], [ES] and [P], assuming $[S] = [S]_0$.

At $t = 0$, $[P] = 0$ and $d[P]/dt = 0$. After some time, $d[P]/dt$ will reach the "steady state" of simple enzyme kinetics. A critical time, t_c, can be defined by extrapolating the linear portion of the [P] curve to intersect the time or horizontal axis (Dixon and Webb, 1964). From the equation of [P] just derived, we have

$$t_c = \{k_1([S]_0 + K_m)\}^{-1}.$$

Note that this critical time depends on the three kinetic constants and also the initial substrate concentration. Together with V_{max} and K_m, t_c provides the third equation to determine the three kinetic constants. Therefore, finally we have:

$$k_1 = 1 / \{t_c([S]_0 + K_m)\},$$

$$k_{-1} = K_m / \{t_c([S]_0 + K_m)\} - V_{max} / [E]_0,$$

and $$k_2 = V_{max} / [E]_0 .$$

For most enzymatic reactions, t_c is usually very small and can only be measured by special devices.

LOW SUBSTRATE CONCENTRATION

In the original paper of Michaelis and Menten (1913), the concentration of the product was measured at various times for different concentrations of the substrate at the beginning of the enzymatic reaction. An example of one of their measurements is illustrated here (Figs. 4-9 to 4-16). For higher initial substrate concentrations, the initial rate of product formation had a constant region followed by another slower constant region. However, for lower initial substrate concentrations, this rate was constant for a very short duration of probably just a few minutes. The reason for this non-linearity was most likely due to the depletion of substrate as time went on. Without this simplifying assumption, the rate equations mentioned above are thus non-linear, and can only be solved numerically.

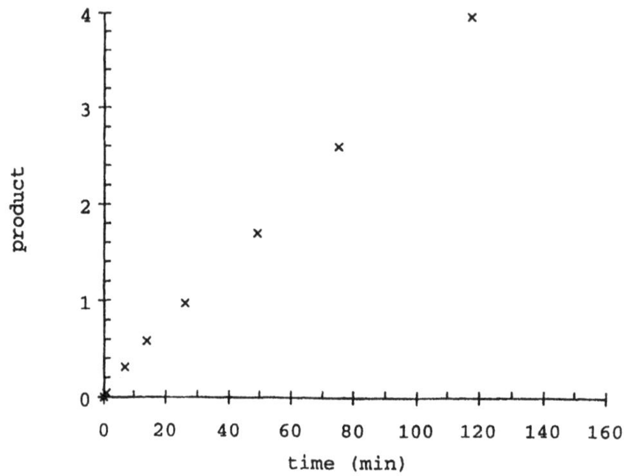

Figure 4-9. Plot of product against time for $[S]_0 = 0.333$ n.

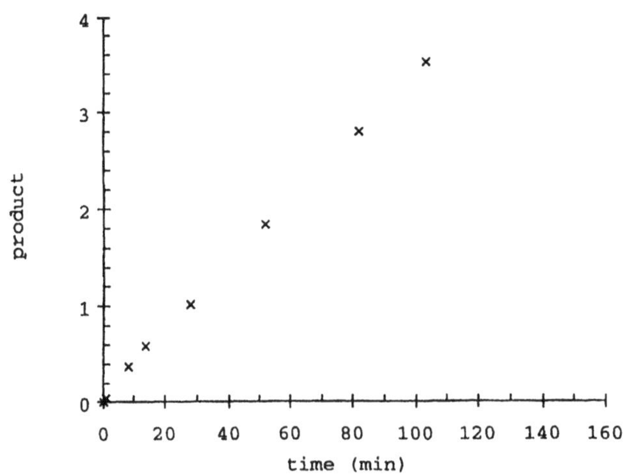

Figure 4-10. Plot of product against time for $[S]_0 = 0.167$ n.

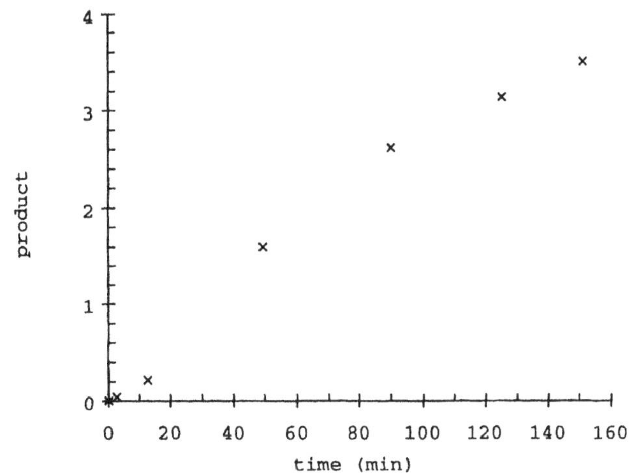

Figure 4-11. Plot of product against time for $[S]_0 = 0.0833$ n, experiment a.

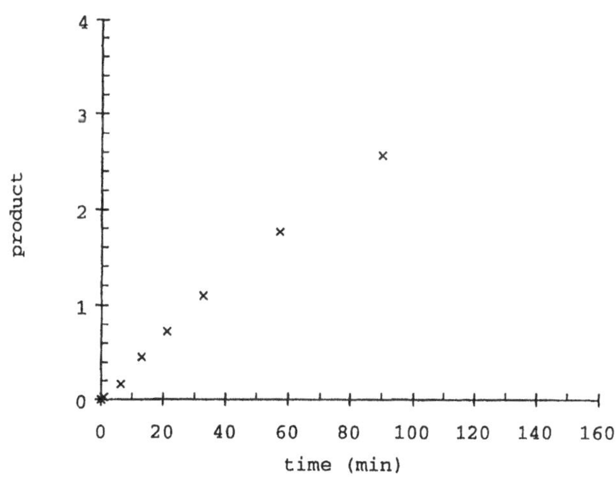

Figure 4-12. Plot of product against time for $[S]_0 = 0.0833$ n, experiment b.

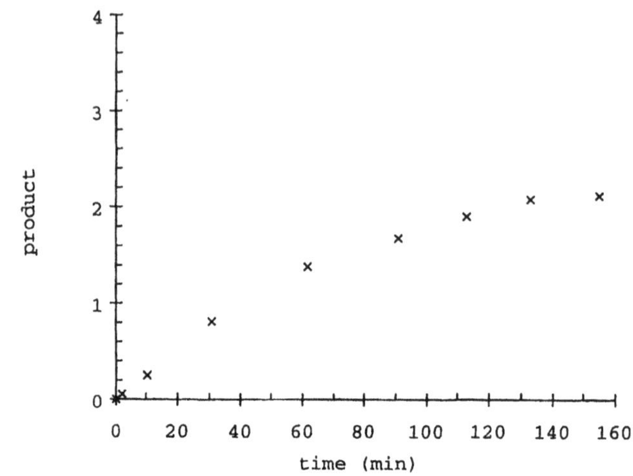

Figure 4-13. Plot of product against time for $[S]_0 = 0.0416$ n.

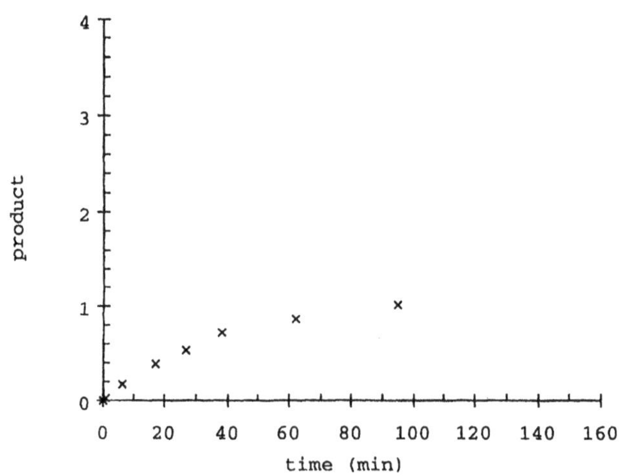

Figure 4-14. Plot of product against time for $[S]_0 = 0.0208$ n.

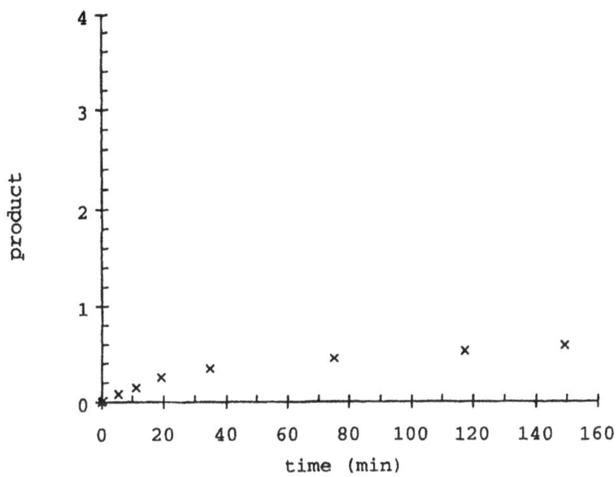

Figure 4-15. Plot of product against time for $[S]_0 = 0.0104$ n.

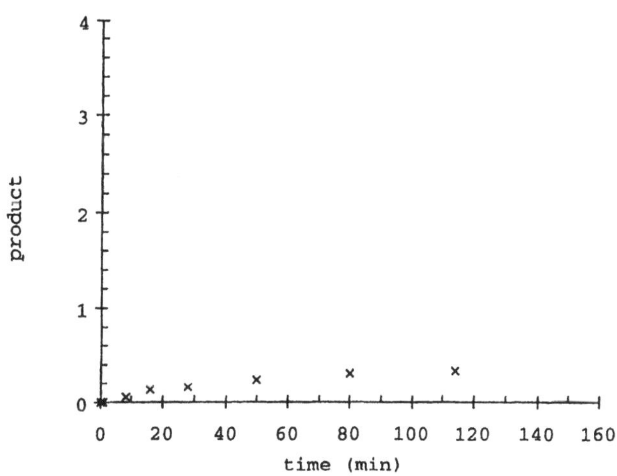

Figure 4-16. Plot of product against time for $[S]_0 = 0.0052$ n.

For high initial substrate concentrations (Figs. 4-9, 4-10), the initial rates of product formation were constant for about 10 minutes and slowed down somewhat. They remained at the slower constant rate for nearly 100 minutes. At intermediate substrate concentration (Figs. 4-11, 4-12), there was a short lag of a few minutes, somewhat indicative of the existence of a transient state discussed above, with a t_c of around 2 to 6 minutes. Beyond 30 minutes, the rates slowed down gradually. For lower substrate concentrations (Figs. 4-13 to 4-16), the initial rates could not be maintained and the rates of product formation leveled off quickly.

NUMERICAL SOLUTION FOR NON-LINEAR RATE EQUATIONS

Since the transition from the initial conditions at t = 0 to the steady states assumed in simple enzyme kinetics occurs very quickly, usually less than one or a few minutes, sometimes in seconds or less, numerical calculations should avoid that time period, or use time increment, Δt, much less than t_c. In the simplest approximation, we have:

$$[P](t+\Delta t) = k_2\,[ES](t)\,\Delta t\,,$$

$$[ES](t+\Delta t) = k_1\,\{[E]_0 - [ES](t)\}\,[S](t)\,\Delta t$$

$$- \{k_{-1} + k_2\}\,[ES](t)\,\Delta t\,,$$

$$[S](t+\Delta t) = -\,k_1\,\{[E]_0 - [ES](t)\}\,[S](t)\,\Delta t$$

and

$$+\,k_2\,[ES](t)\,\Delta t\,.$$

In this situation, [S] is no longer considered as a constant. It will gradually reduce in concentration. As a result, [ES] will also gradually decrease. Eventually, the substrate will be exhausted, and [ES] will also approach zero. Schematically, the time courses of [S], [E], [ES] and [P] are illustrated in Fig. 4-17. The rate of product generation eventually levels off, as actually noticed by Michaelis and Menten (1913) for low initial substrate concentrations (Fig. 4-15, 4-16).

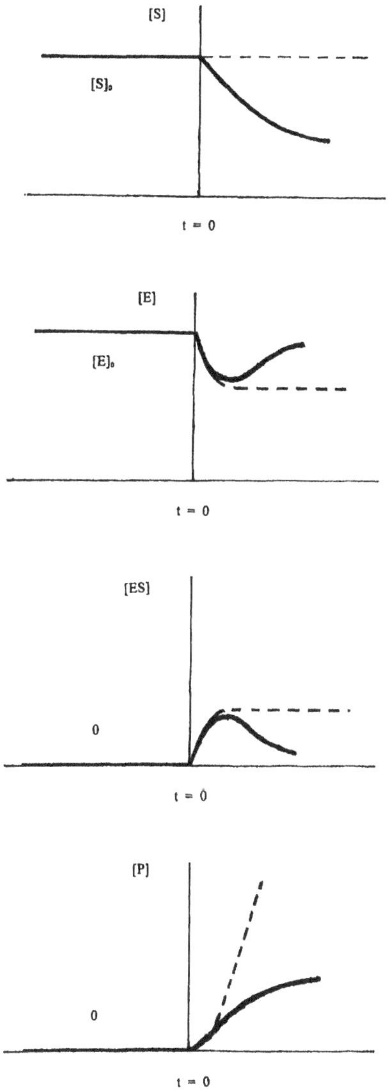

Figure 4-17. Schematic time courses of [S], [E], [ES] and [P] for low $[S]_0$.

Other methods for more accurate numerical integration of non-linear ordinary differential equations are commonly used, e.g. Taylor-series, Euler, Runge-Kutta, etc. (see, for example, Gerald, 1978).

INHIBITORS

There are two simply types of enzyme inhibitors: competitive and non-competitive (Dixon and Webb, 1964). Other more complicated types are also possible. In the case of the fully competitive inhibitor, we have:

$$E + S \underset{k_{-1}}{\overset{k_1}{\rightleftharpoons}} ES \xrightarrow{k_2} E + P,$$

and

$$E + I \underset{k_{-3}}{\overset{k_3}{\rightleftharpoons}} EI,$$

where I denotes the inhibitor. If we make the similar assumptions as in the case of simple enzyme kinetics, i.e. $d[ES]/dt = 0$, $d[EI]/dt = 0$, $[S] = [S]_0$, and $[I] = [I]_0$, we have:

$$k_1 [E] [S]_0 - (k_{-1} + k_2) [ES] = 0,$$

and

$$k_3 [E] [I]_0 - k_{-3} [EI] = 0,$$

so that

$$[E] \{1 + k_1 [S]_0 / (k_{-1} + k_2) + k_3 [I]_0 / k_{-3}\} = [E].$$

Since $$V = d[P]/dt = k_2[ES],$$

we have $$V = V_{max}[S]_0/\{[S]_0 + K_m + K_m[I]_0/K_i\},$$

where $K_i = k_{-3}/k_3$. In the presence of various amounts of the inhibitor, V_{max} does not change while the apparent K_m defined as:

$$\text{apparent } K_m = K_m + K_m[I]_0/K_i$$

will be increased as $[I]_0$ increases. Usually, $1/V$ is plotted against $1/[S]_0$ as shown in Fig. 4-18.

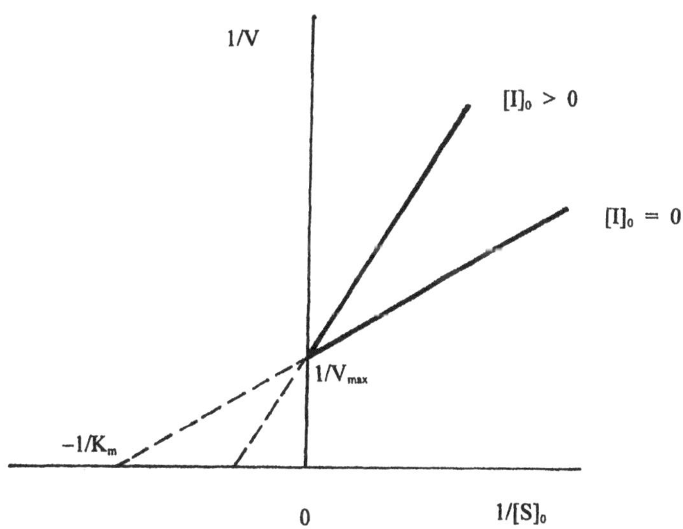

Fig. 4-18. Fully competitive inhibitor.

The simplest type of non-competitive inhibitor can be considered as:

$$E + S \underset{k_{-1}}{\overset{k_1}{\rightleftharpoons}} ES \xrightarrow{k_2} E + P\,,$$

$$E + I \underset{k_{-3}}{\overset{k_3}{\rightleftharpoons}} EI\,,$$

$$ES + I \underset{k_{-3}}{\overset{k_3}{\rightleftharpoons}} ESI\,,$$

and

$$EI + S \underset{k_{-1}}{\overset{k_1}{\rightleftharpoons}} ESI\,.$$

The resulting equations, even with the assumptions similar to simple enzyme kinetics, are quite complicated. With the requirement of conserving enzyme molecules, i.e.

$$[E] + [ES] + [EI] + [ESI] = [E]_0\,,$$

the three unknown concentrations, [ES], [EI] and [ESI], satisfy the following set of simultaneous algebraic equations:

$$\{k_1[S]_0 + k_3[I]_0 + k_{-1} + k_2\}\ [ES] + k_1[S]_0\ [EI] + \{k_1[S]_0 - k_{-3}\}\ [ESI]$$

$$= k_1[E]_0[S]_0\ ,$$

$$k_3[I]_0\ [ES] + \{k_3[I]_0 + k_1[S]_0 + k_{-3}\}\ [EI] + \{k_3[I]_0 - k_{-1}\}\ [ESI]$$

$$= k_3[E]_0[I]_0\ ,$$

and $$k_3[I]_0\ [ES] + k_1[S]_0\ [EI] - \{k_{-1} + k_{-3}\}\ [ESI] = 0\ .$$

Using the formula given in the Appendix, we can solve for [ES] to give:

$$[ES] = \frac{\begin{vmatrix} [E]_0[S]_0 & [S]_0 & [S]_0 - K_i/\alpha \\ [E]_0[I]_0 & [I]_0 + \alpha[S]_0 + K_i & [I]_0 - \alpha K_m \\ 0 & \alpha[S]_0 & -\{\alpha K_m + K_i\} \end{vmatrix}}{\begin{vmatrix} [S]_0 + [I]_0/\alpha + K_m & [S]_0 & [S]_0 - K_i/\alpha \\ [I]_0 & [I]_0 + \alpha[S]_0 + K_i & [I]_0 - \alpha K_m \\ [I]_0 & \alpha[S]_0 & -\{\alpha K_m + K_i\} \end{vmatrix}}$$

where $K_m = (k_{-1} + k_2)/k_1$ as before, $K_i = k_{-3}/k_3$ and $\alpha = k_1/k_3$. On simplying the messy algebra, we have:

$$[ES] = \frac{[E]_0\ [S]_0\ K_i}{\{[S]_0 + K_m\}\ \{[I]_0 + K_i\}}\ .$$

In this case, the velocity of the enzymatic reaction is:

$$V = \frac{V_{max}\,[S]_0}{\{[S]_0 + K_m\}\;\{1 + [I]_0/K_i\}}.$$

The K_m does not change while the apparent V_{max} as defined

$$\text{apparent } V_{max} = V_{max} / \{1 + [I]_0/K_i\}$$

is reduced as $[I]_0$ increases. The $1/V$ vs. $1/[S]_0$ plot for non-competitive inhibitor is shown in Fig. 4-19.

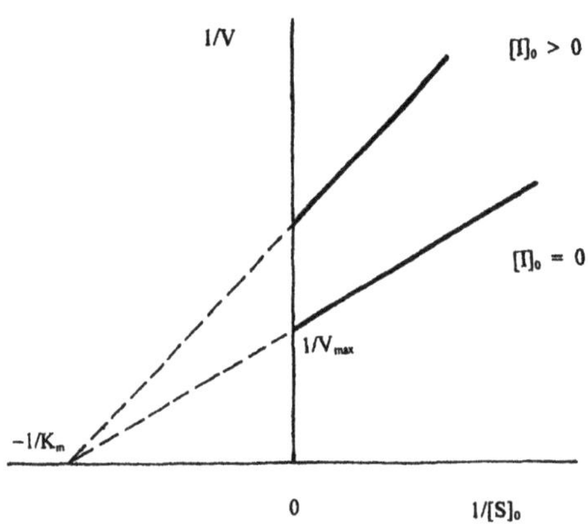

Figure. 4-19. Non-competitive inhibitor.

Many of the new drugs developed during recent decades are enzyme inhibitors of various types.

FUTURE STUDIES

As illustrated here, we have only discussed about the very simplified kinetic reaction of a single enzyme. This formulation can be incorporated into the study of time-dependent kinetics of hemoglobin saturation (see, for example, Farmery and Roe, 1996), the equilibrium states of which has been discussed in Chapter 2. It can also be used as a building block for analyzing metabolic pathways involving a collection of different enzymes (see, for example, Yang *et al.*, 1999). However, in most cases, only the K_m and V_{max} are of interest to most biochemists and molecular biologists.

Some of the enzymatic reactions are very fast. Special devices have been developed to measure such reactions in the millisecond range (see, for example, Cornish-Bowden, 1979). Recently, such measurements are gradually becoming more routine (see, for example, Sclavi *et al.*, 1998; Koltermann *et al.*, 1998). Then, the basic mechanism of how enzymes function at the molecular level can be deciphered.

REFERENCES

Cornish-Bowden A (1979) *Fundamentals of Enzyme Kinetics.* Butterworths London.

Dixon M and Webb EC (1964) *Enzymes.* Academic Press Inc., Publishers, New York.

Farmery AD and Roe PG (1996) A model to describe the rate of oxyhaemoglobin desaturation during apnoea. *Br. J. Anaesth.*, **76**, 284-291.

Gerald CF (1978) *Applied Numerical Analysis.* Addison-Wesley Publishing Company, Reading, MA.

Koltermann A, Kettling U, Bieschke J, Winkler T and Eigen M (1998) Rapid assay processing by integration of dual-color fluorescence cross-correlation spectroscopy: high throughput screening for enzyme activity. *Proc. Nat. Acad. Sci. USA*, **95**, 1421-1426.

McGregor WG, Phillips J and Suelter CH (1974) Purification and kinetic characterization of a monovalent cation-activated glycerol dehydrogenase from *Aerobacter aerogenes. J. Biol. Chem.*, **249**, 3132-3139.

Michaelis L and Menten ML (1913) Die Kinetik der Invertinwirkung. *Biochem. Z.*, **49**, 333-369.

Sclavi B, Sullivan M, Chance MR, Brenowitz M and Woodson SA (1998) RNA folding at millisecond intervals by synchrotron hydroxyl radial footprinting. *Science*, **279**, 1940-1943.

Yang YT, San KY and Bennett GN (1999) Redistribution of metabolic fluxes in *Escherichia coli* with fermentative lactate dehydrogenase overexpression and deletion. *Metab. Eng.*, **1**, 141-152.

EXERCISE

Write down the rate equations for two consecutive enzymes, alcohol dehydrogenase and aldehyde dehydrogenase, for the metabolism of ethanol, assuming simple enzyme kinetics for both.

CHAPTER 5

THE ϕ AND ψ ANGLES OF PROTEINS

INTRODUCTION

Most proteins are liner polymers of 20 different amino acid residues, e.g. myoglobin discussed in Chapter 2. Hemoglobin consists of four chains, two α and two β, which are held together by hydrophobic interactions. Other proteins, e.g. antibodies discussed in Chapter 1, have these linear polymers connected by covalent bonds.

The amino acid residues are joined together by peptide bonds usually in the trans-configuration (Fig. 5-1). Due to the de-localization of electrons, that bond is considered as "partially" double. As a result, locally the peptide bond is more or less flat, i.e. the two neighboring C_α, N, H, C and O atoms shown in Fig. 5-1 all lie in one plane.

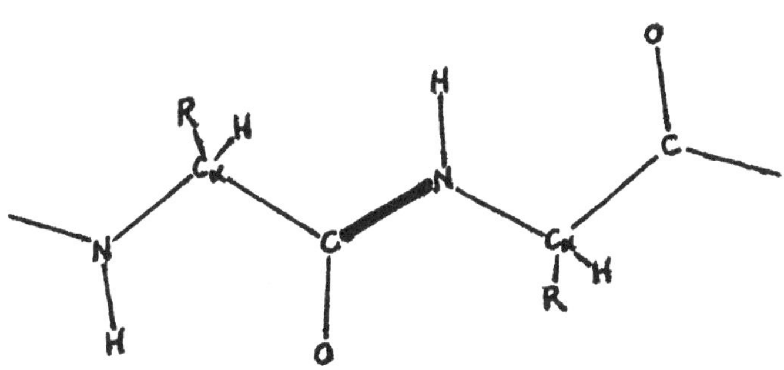

Figure 5-1. The peptide bond.

For each amino acid residue, it is bracketed by two peptide bonds, one on the N-side and the other on the C-side. Since there is no rotational degree of freedom for peptide bonds, the protein backbone contains two single bonds for each residue, free to rotate. They are the N–C_α and the C_α–C bonds (Fig. 5-2). These rotations are designated as ϕ and ψ respectively. They will be defined precisely.

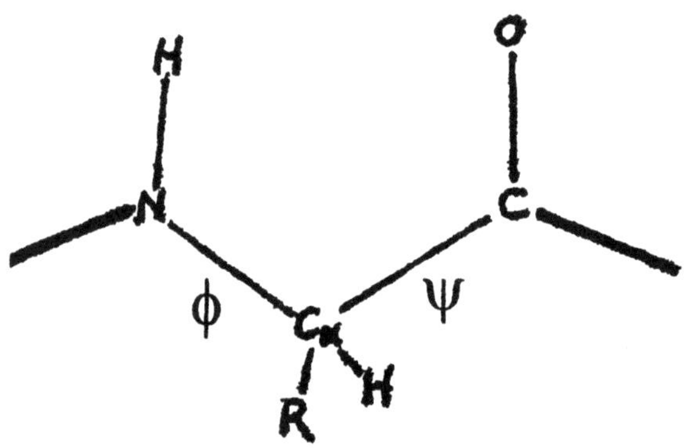

Figure 5-2. Two single bonds free to rotate in an amino acid residue.

THE ϕ AND ψ ANGLES

The best method of displaying the ϕ and ψ angles in three dimension is to make use of the paper model illustrated in Fig. 5-3, as originally used by Pauling to discover the α–helix. Fig. 5-2 is duplicated, and the N–C_α and the C_α–C bonds are extended beyond the C_α–atom. Cut along the heavy lines drawn on the page. This includes the triangle outlined by the above extensions. Fold along the N–C_α bond, as well as the C_α–C bond. These folds represent the ϕ and ψ angles respectively. The ϕ angle is the angle between the plane defined by C_{-1}–N–C_α and the plane defined by N–C_α–C.

Similarly, the ψ angle is the angle between the plane defined by N–C_α–C and the plane defined by C_α–C–N_{+1}.

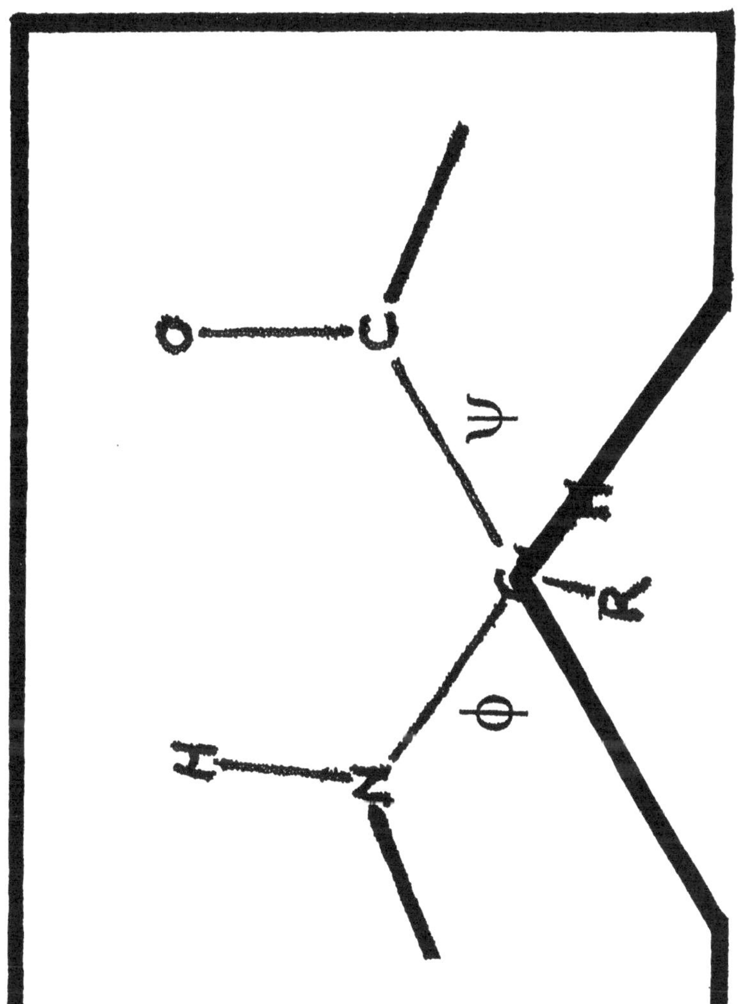

Figure 5-3. Paper model illustrating the ϕ and ψ angles.
(Make a copy and cut along the heavy lines.)

PRECISE DEFINITION OF THE ϕ ANGLE

To define the ϕ angle precisely, look down the N–C_α bond (Fig. 5-4). Hold the C_α–atom fixed, together with C, H and the R-group. Rotate the N, H and C_{-1} atoms counter-clockwise. That defined the ϕ angle. Unfortunately, there are two conventions. The old one used in some textbooks (see, for example, Dickerson and Geis, 1969) defines the ϕ angle as 0° in the flat extended configuration. The new one accepted by current literatures defines ϕ angle as −180° in the same configuration.

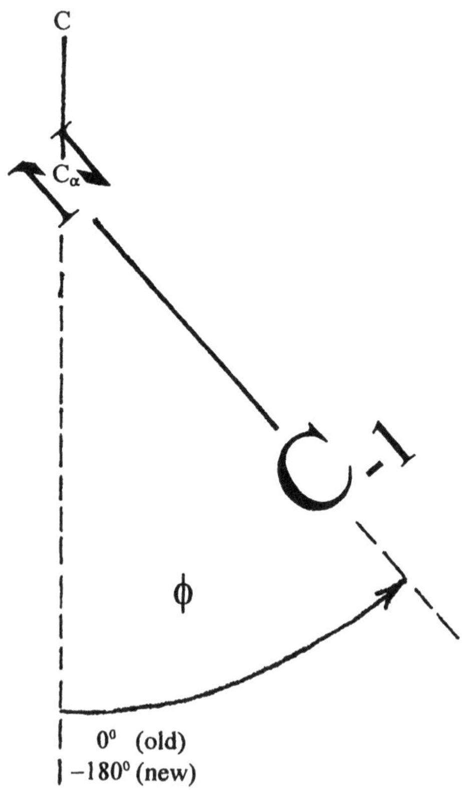

Figure 5-4. Definition of the ϕ angle.

PRECISE DEFINITION OF THE ψ ANGLE

Similarly, to define the ψ angle precisely, look down the C–C_α bond (Fig. 5-5). Again hold the C_α–atom fixed, together with N, H and the R-group. Rotate the C, O and N_{+1} atoms counter-clockwise. That defines the ψ angle. Also, there are two conventions identical to the ϕ angle.

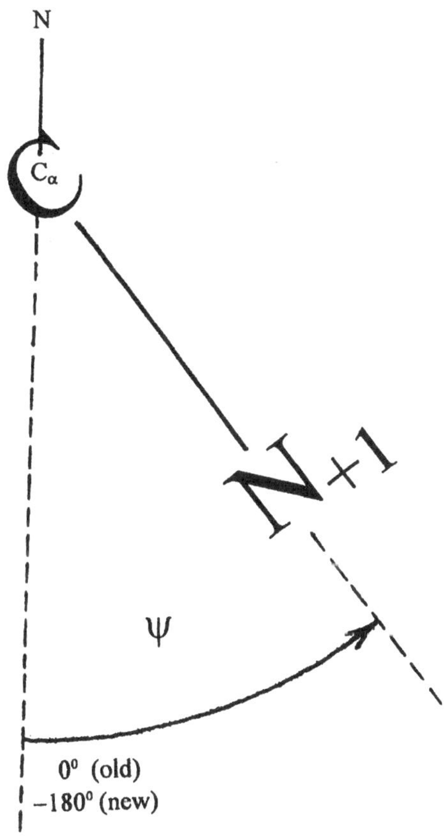

Figure 5-5. Definition of the ψ angle.

The old and new conventions for both the ϕ and ψ angles are illustrated in Table 5-1.

Table 5-1. Old and new conventions for the ϕ and ψ angles.

Old	New
0^0	-180^0
30^0	-150^0
60^0	-120^0
90^0	-90^0
120^0	-60^0
150^0	-30^0
180^0	0^0
210^0	$+30^0$
240^0	$+60^0$
270^0	$+90^0$
300^0	$+120^0$
330^0	$+150^0$
360^0	$+180^0$

Sometimes, however, in the new convention, the + sign may be omitted resulting in ambiguity.

PROTEIN DATA BANK

Since the determination of the three dimensional structures of myoglobin and hemoglobin as previously mentioned in Chapter 2, many other three dimensional structures of numerous proteins have been analyzed by X-ray diffraction studies, NMR measurements, theoretical predictions, etc. Most of these structures are available from the Protein Data Bank at website http://www.rcsb.org/pdb/, except possibly those predicted theoretically. The Cartesian coordinates for non-hydrogen atoms are usually reported. However, ϕ and ψ angles are not available. In this Chapter, how these angles can be calculated from atomic coordinates will be explained in detail.

UNIT NORMALS TO PLANES

To define a plane, we need to specify its unit normal. For the three planes shown in Fig. 5-3, C_{-1}–N–C_α , N–C_α–C and C_α–C–N_{+1} , the unit normals can be calculated from the vectors drawn from C_α to C_{-1} , N, C and N_{+1} . Therefore, starting with the atomic coordinates of C_{-1}, N, C_α , C and N_{+1} , we need to translated the origin of the coordinate system to C_α . Then, in the translated system, the coordinates for C_{-1} , N, C and N_{+1} specify the vectors $\mathbf{C}_{-1}$, **N**, **C** and $\mathbf{N}_{+1}$ respectively. The unit normals for the three planes are as follows:

plane C_{-1}–N–C_α:

$$\mathbf{U} = \frac{\mathbf{N} \times \mathbf{C}_{-1}}{|\mathbf{N} \times \mathbf{C}_{-1}|},$$

plane N–C_α–C:

$$\mathbf{V} = \frac{\mathbf{C} \times \mathbf{N}}{|\mathbf{C} \times \mathbf{N}|}.$$

and plane C_α–C–N_{+1}:

$$\mathbf{W} = \frac{\mathbf{N}_{+1} \times \mathbf{C}}{|\mathbf{N}_{+1} \times \mathbf{C}|}.$$

ϕ AND ψ ANGLES AS ANGLES BETWEEN PLANES

Since the ϕ angle is the angle between planes C_{-1}–N–C_α and N–C_α–C, we have:

$$\cos\phi = \mathbf{V} \cdot \mathbf{U}.$$

However, whether ϕ is positive or negative cannot be specified. A triple product is needed to determine the sign of ϕ:

$$\text{if} \qquad \mathbf{C} \times \mathbf{N} \cdot \mathbf{C}_{-1} > 0, \qquad \text{then} \qquad \phi > 0;$$

$$\text{if} \qquad \mathbf{C} \times \mathbf{N} \cdot \mathbf{C}_{-1} < 0, \qquad \text{then} \qquad \phi < 0.$$

The ϕ angle thus calculated is in the old convention, if positive. If negative, add 360° to get the value in the old convention. To express ϕ in the new convention, subtract 180° from that value.

Similarly, since the ψ angle is the angle between planes N–C_α–C and C_α–C–N_{+1}, we have:

$$\cos \psi = \mathbf{V} \cdot \mathbf{W}.$$

Again, a triple product is needed to determine the sign of ψ:

$$\text{if} \qquad \mathbf{C} \times \mathbf{N} \cdot \mathbf{N}_{+1} > 0, \qquad \text{then} \qquad \psi < 0;$$

$$\text{if} \qquad \mathbf{C} \times \mathbf{N} \cdot \mathbf{N}_{+1} < 0, \qquad \text{then} \qquad \psi > 0.$$

Also, the ψ angle thus calculated is in the old convention, if positive. If negative, add 360° to get the value in the old convention. To express ψ in the new convention, subtract 180° from that value.

AN EXAMPLE

Consider a typical set of Cartesian coordinates for non-hydrogen atoms C_{-1}, N, C_α, C and N_{+1} as listed in Table 5-2.

Table 5-2. A typical set of Cartesian coordinates for atoms C_{-1}, N, C_α, C and N_{+1}.

	x	y	z
C_{-1}	-13.351	10.195	6.742
N	-14.244	9.389	7.285
C_α	-14.271	9.077	8.721
C	-13.756	7.654	8.943
N_{+1}	-12.711	7.340	8.199

They are in Angstrom units, A, or 0.1 nm. The third decimal place may not be reliable. As discussed later, the first two decimal places are generated by the use of a standard peptide unit. To position the C_α–atom at the origin, the above set of coordinates is translated to give the result listed in Table 5-3:

Table 5-3. Translated coordinates of atoms listed in Table 5-2, with C_α–atom positioned at the origin of the new coordinates.

	x	y	z
C_{-1}	0.920	1.118	-1.979
N	0.027	0.312	-1.436
C_α	0.0	0.0	0.0
C	0.515	-1.423	0.222
N_{+1}	1.560	-1.737	-0.522

The vectors $\mathbf{C}_{-1}$, **N**, **C** and $\mathbf{N}_{+1}$ are thus defined by the new coordinates (Fig. 5-6). From these, we can calculate the unit normals to the three planes (Fig. 5-3), **U**, **V** and **W**, as mentioned above. We have:

$$\mathbf{C} \times \mathbf{N} = 1.974\,\mathbf{i} + 0.746\,\mathbf{j} + 0.199\,\mathbf{k},$$

where **i**, **j** and **k** are unit vectors in the x-, y- and z-directions.

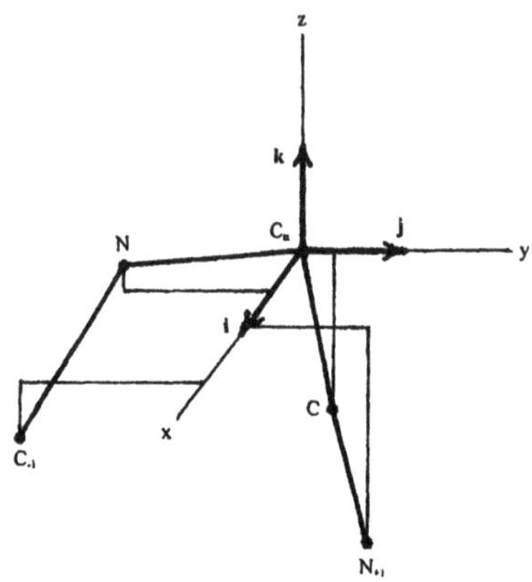

Figure 5-6. Locations of C_{-1} , C, N and N_{+1} in the coordinate system where C_α is at the origin.

Therefore, $$|\mathbf{C} \times \mathbf{N}| = 2.120 ,$$

so that $$\mathbf{U} = 0.931\,\mathbf{i} + 0.352\,\mathbf{j} + 0.094\,\mathbf{k} .$$

Similarly, we can calculate:

$$\mathbf{V} = 0.557\,\mathbf{i} - 0.818\,\mathbf{j} - 0.145\,\mathbf{k} ,$$

and $$\mathbf{W} = -0.611\,\mathbf{i} - 0.333\,\mathbf{j} - 0.718\,\mathbf{k} .$$

The unit vectors **U**, **V** and **W** are shown in Fig. 5-7.

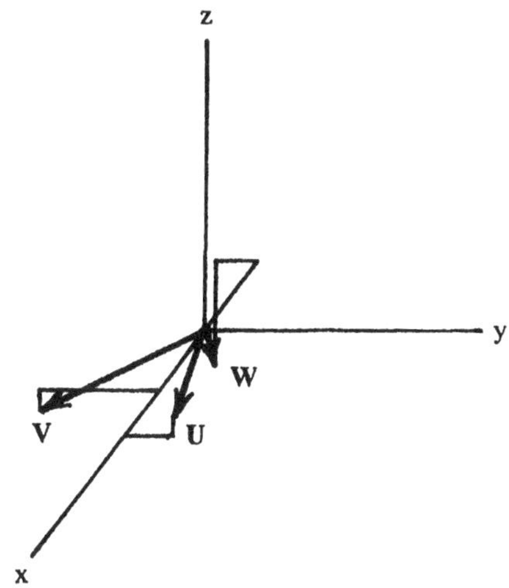

Figure 5-7. Unit vectors **U**, **V** and **W** to the three planes illustrated in Fig. 5-3.

Since $$\cos\phi = \mathbf{U}\cdot\mathbf{V} = 0.217\,,$$

so $$\phi = 77^{o}\,.$$

To determine its sign, we calculate the triple product

$$\mathbf{C} \times \mathbf{N}\cdot\mathbf{C}_{-1} = 2.256\,.$$

Therefore, the value of ϕ is positive 77° in the old convention. In the new convention,

$$\phi = 77^{o} - 180^{o} = -103^{o}\,.$$

Similarly, $\cos\psi = \mathbf{U}\cdot\mathbf{W} = -0.754$.

So $\psi = 139^0$.

Then we calculate the triple product

$$\mathbf{C} \times \mathbf{N}\cdot\mathbf{N}_{+1} = 1.680 .$$

Therefore, in the old convention,

$$\psi = -139^0 + 360^0 = 221^0 ;$$

and in the new convention,

$$\psi = 221^0 - 180^0 = +41^0 .$$

The plus sign is very important in the new convention, in order not to be confused with the old convention.

BOND DISTANCES AND BOND ANGLES

The above set of atomic coordinates can also be used to calculate the bond distances and bond angles of the peptide backbone. The bond distances are:

$C_{-1} - N$	1.32 A
$N - C_\alpha$	1.47 A
$C_\alpha - C$	1.53 A

$C - N_{+1}$ 1.32 A

The bond angles are:

$C_{-1} - N - C_{\alpha}$ 123^{0}

$N - C_{\alpha} - C$ 109^{0}

$C_{a} - C - N_{+1}$ 114^{0}

As discussed later, these values are very similar to those for a standard peptide unit. In fact, since most of the X-ray diffraction studies of protein three-dimensional structures do not have enough resolution to locate the positions of atoms, the standard peptide unit is used to construct the final atomic model.

RAMACHANDRAN (ϕ,ψ) PLOT AND STERIC HINDRANCE

Based on quantum mechanical calculations (see, for example, Schiff, 1955), electron density distributions around hydrogen atoms are relatively diffused. Therefore, hydrogen atoms are usually not "visible" in X-ray diffraction studies. However, based on hard-sphere approximation or various empirical interactions between non-bounded atoms, hydrogen atoms play an extremely important role in determining whether a certain configuration is sterically allowed. Ramakrishnan and Ramachandran (1965) made an exhaustive study of such configurations for amino acid residues, based on ϕ and ψ angles as defined above. They introduced the Ramachandran (ϕ,ψ) plot, although they used an angle designated as ϕ' which was different from ψ. Hard-spheres are usually assumed for non-bounded atoms, with a minimum contact distance allowed between any two atoms as listed in Table 5-4.

Table 5-4. Empirical minimum distances between non-bounded atoms in Angstrom units, A.

Non-bounded atoms	Normal limit	Extreme limit
C----C	3.0	2.9
C----H	2.4	2.2
C----N	2.9	2.8
C----O	2.8	2.7
H----H	2.0	1.9
H----N	2.4	2.2
H----O	2.4	2.2
N----N	2.7	2.6
N----O	2.7	2.6
O----O	2.7	2.6

To calculate steric hindrance, the bond distances and bond angles of a standard peptide unit must also be given, as shown in Fig. 5-8. The bond distance from H to N or C is assumed to be around 1.05 A.

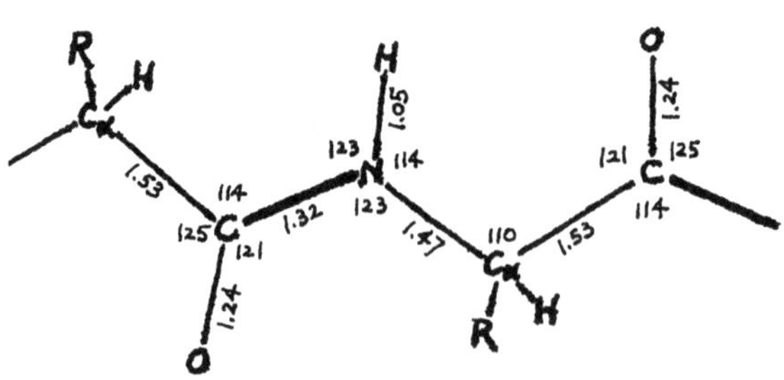

Figure 5-8. Bond angles and bond distances of a standard peptide unit.

For example, consider {0°,0°) in the new convention (Fig. 5-9). The O bounded to C_{-1} and the H bounded to N_{+1} are separated by less than 0.2 A (< 2.2 A). This configuration is thus sterically hindered.

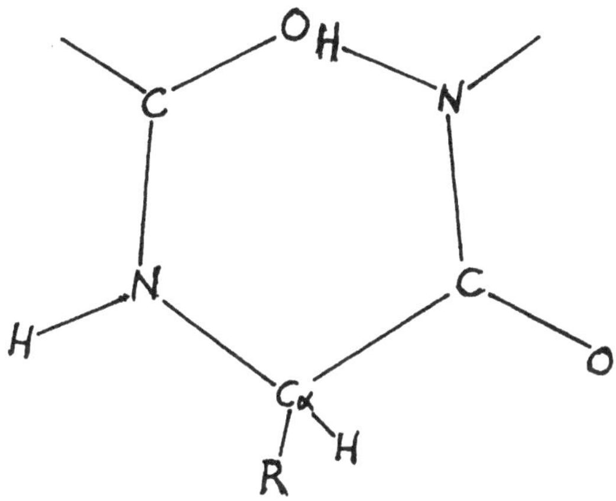

Figure 5-9. Configuration for (0°,0°) in the new convention.

Regardless of whether the old or new convention is used for ϕ and ψ angles, the same Ramanchandran (ϕ,ψ) plot is used by all researchers for ranges of 0° to 360° in the old convention and −180° to +180° in the new convention (Fig. 5-10). The configurations represented by (0°,0°) and (−103°,+41°) in the new convention are plotted. The point represented by (0°,0°) has bad steric hindrance as mentioned above. The other point, (−103°,+41°), presumably should be sterically allowed, since it is found in one of the structures of a protein.

The steric hindrance calculations used by Ramakrishnan and Ramachandran (1965) are, however, totally independent of X-ray diffraction studies of proteins. Thus, they provide a different aspect for the study of protein configurations. The distances between any two non-bounded atoms can be calculated from the (ϕ,ψ) angles and the bound distances and angles of a standard peptide, and compared with the minimum contact distances listed in Table 5-4 as to whether they are allowed.

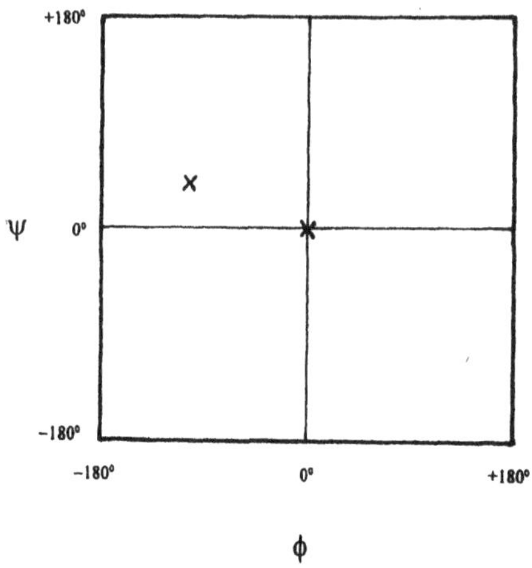

Figure 5-10. The Ramanchandran (ϕ,ψ) plot.

Such calculations will depend on the chemical structure of the amino acid residue.

ALLOWED CONFIGURATIONS OF AMINO ACID RESIDUES

For the 20 different amino acid residues, they can be divided into three groups. Glycine residue is different from the others, with its R-group attached to C_α–atom being a hydrogen atom. Proline residue has a ring structure for its side chain joining to the backbone N–atom. Sometimes, it can be in the cis-configuration. All the other 18 amino acid residues have a C_β–atom covalently bonded to the C_α–atom. They are usually considered together for steric hindrance calculations, and represented by the alanine residue. Ramakrishnan and Ramachandran (1965) have calculated the allowed regions for alanine and glycine residues as shown in Fig. 5-11 and Fig. 5-12 respectively. The extreme minimal limits are used for these plots. In the case of proline residue, the ring structure of its side chain restricts its ϕ angle to about -60^0 in the new convention (Fig. 5-13). Proline residues may sometimes be in the *cis*-configuration, and have ϕ around $+60^0$ instead.

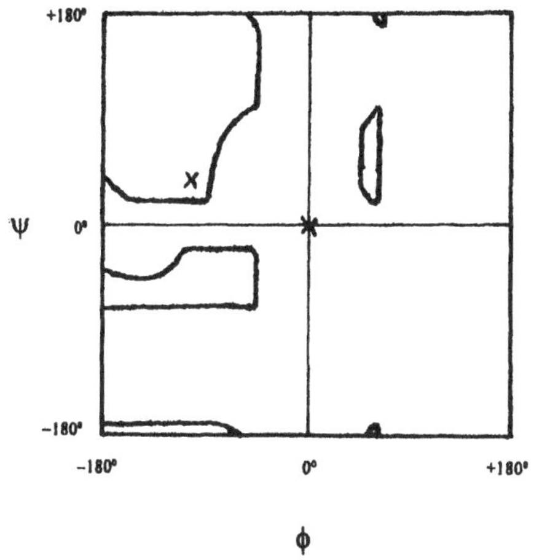

Figure 5-11. Allowed region in the Ramachandran plot for alanine residue.

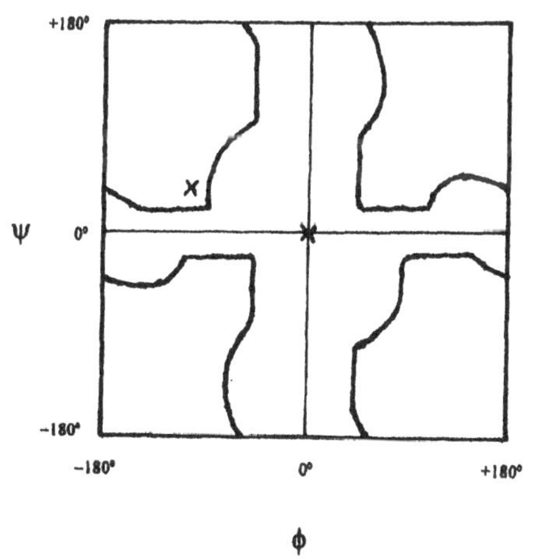

Figure 5-12. Allowed region in the Ramachandran plot for glycine residue.

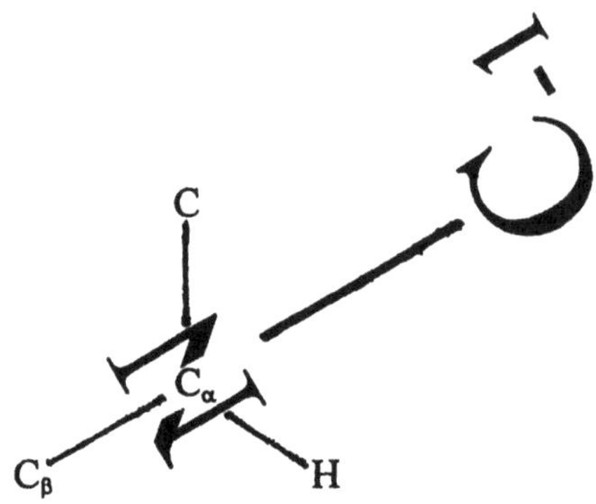

Figure 5-13. Restriction of ϕ to about -60^0 for proline residue.

The calculations for the allowed regions are very sensitive to the value of the τ or the N–C_α–C angle. This angle is usually assumed to be 110^0, but may indeed vary by about 5^0 (Ramakrishnan and Ramachandran, 1965).

In general, for every three-dimensional structure of protein determined by X-ray diffraction studies, the (ϕ,ψ) angles should be calculated from the atomic coordinates of non-hydrogen atoms, as discussed above, in order to investigate how the three-dimensional structure can be refined. However, these (ϕ,ψ) angles are usually not listed in the protein data bank, http://www.rcsb.org/pdb/. In a recent study of the structure of 30S ribosomal subunits, Wimberly *et al.* (2000) mentioned:

"For the proteins, 95.7% of the residues were in the core or allowed regions of the Ramachandran plot, 2.4% in the generously allowed region and 1.9% in the disallowed region."

The configurations of the 1.9% of disallowed amino acid residues may have their (ϕ,ψ) angles adjusted to be included in the allowed regions after refinement.

CALCULATIONS OF COORDINATES OF NON-BONDING ATOMS

In a three-dimensional Cartesian coordinate system as shown in Fig. 5-14, place the C_α–atom at the origin, the N–atom on the negative y-axis, and the C–atom in the y,z-plane.

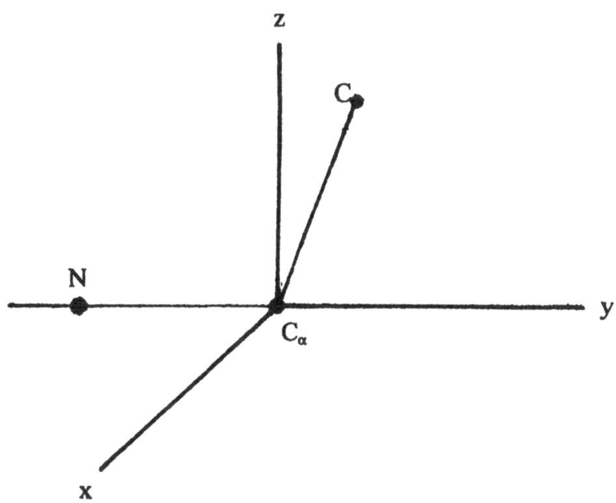

Figure 5-14. Locations of protein backbone atoms.

Thus, the coordinates of these three atoms are:

C_α (0,0,0) ,

N (0,-1.47,0) , and

C (0,0.52,1.44) .

The y-coordinate for C–atom is calculated from:

$$1.53\cos(180^\circ - 110^\circ) = 1.53\cos 70^\circ$$

$$= 0.52\,,$$

and its z-coordinate calculated as:

$$1.53\sin 70^\circ = 1.44\,.$$

With the use of the rotation matrix discussed in the Appendix and the symmetrical locations of atoms bounded to the C_α–atom, the coordinates of the H–atom and the C_β–atom can be calculated as follows. The y-coordinate for the H–atom is:

$$1.05\cos 70^\circ = 0.36\,.$$

The x-coordinate for the H–atom is:

$$1.05\sin 70^\circ\cos(120^\circ - 90^\circ) = 1.05\sin 70^\circ\cos 30^\circ$$

$$= 0.86\,.$$

Also, the z-coordinate for the H–atom is:

$$-1.05\sin 70^\circ\sin 30^\circ = -0.49\,.$$

Therefore, we have approximately the coordinates of the H–atom as:

$H(C_\alpha)$ $(0.86,0.36,-0.49)$.

Consider the bond distance between C_α– and C_β–atoms to be 1.53 A. Then a similar calculation gives the coordinates of the C_β–atom as:

$C_\beta(C_\alpha)$ $(-1.25,0.52,-0.72)$.

The coordinates of other atoms will depend on the $(\phi.\psi)$ angles. Fig. 5-14 shows the protein backbone structure in a stretched out configuration corresponding to $(0^o,0^o)$ in the old convention. Thus, coordinates of various atoms will first be given in the old convention, and then repeated in the new convention. Since

$$\sin(\theta - 180^o) = -\sin\theta ,$$

and

$$\cos(\theta - 180^o) = -\cos\theta ,$$

the expressions in the old and new conventions differ by a sign change, whenever ϕ or ψ is involved.

With the use of rotation matrix discussed in the Appendix, those depend on the ϕ angle are:

H (N) $(-0.96 \sin\phi,-1.90,0.96 \cos\phi)$ (OLD)

$(0.96 \sin\phi,-1.90,-0.96 \cos\phi)$ (NEW)

C_{-1} $(1.11 \sin\phi,-2.19,-1.11 \cos\phi)$ (OLD)

$(-1.11 \sin\phi,-2.19,1.11 \cos\phi)$ (NEW)

O (C_{-1}) $(2.26 \sin\phi, -1.73, -2.26 \cos\phi)$ (OLD)

$(-2.26 \sin\phi, -1.73, 2.26 \cos\phi)$ (NEW)

Those depend on the ψ angle are:

O (C) $(1.06 \sin\psi,$

$-1.06 \cos\psi \cos 20^{o} + 2.17 \sin 20^{o},$

$1.06 \cos\psi \sin 20^{o} + 2.17 \cos 20^{o})$ (OLD)

$(-1.06 \sin\psi,$

$1.06 \cos\psi \cos 20^{o} + 2.17 \sin 20^{o},$

$-1.06 \cos\psi \sin 20^{o} + 2.17 \cos 20^{o})$ (NEW)

N_{+1} $(-1.21 \sin\psi,$

$1.21 \cos\psi \cos 20^{o} + 2.07 \sin 20^{o},$

$-1.21 \cos\psi \sin 20^{o} + 2.07 \cos 20^{o})$ (OLD)

$(1.21 \sin\psi,$

$-1.21 \cos\psi \cos 20^{o} + 2.07 \sin 20^{o},$

$1.21 \cos\psi \cos 20^{o} + 2.07 \cos 20^{o})$ (NEW)

H (N_{+1}) (–2.09 sinψ,

2.09 cosψ cos20° + 1.50 sin20°,

–2.09 cosψ sin20° + 1.50 cos20°) (OLD)

(2.09 sinψ,

–2.09 cosψ cos20° + 1.50 sin20°,

2.09 cosψ sin20° + 1.50 cos20°) (NEW)

Since the minimum allowable distances between non-bounded atoms are usually accurate only to the first decimal place (table 5-4), some error in the second decimal place of the above coordinates can be tolerated. In general, such hard-sphere approximations give reasonable good result.

OTHER INTERACTION POTENTIALS BETWEEN NON-BOUNDED ATOMS

The interaction between non-bounded atoms can be calculated theoretically by dipole-dipole interactions (Schiff, 1955) for relatively large separations. This gives an attraction between non-bounded atoms proportional to the six power of their separation For interactions within short distances, various empirical potentials have been suggested. One of the most commonly used potentials to study protein folding is the Lennard-Jones potential (see, for example, Fogolari *et al.*, 1996; Clementi *et al.*, 1999). In this case, there is a repulsion between non-bounded atoms proportional to the twelve power of their separation. Sometimes, the Buckingham potential (see, for example, White, 1997) is also used. The repulsion is considered to be exponential. Allowed regions in the (ϕ,ψ) plot (Brant *et al.*, 1967) can also be calculated using these potentials (Fig. 5-15, 5-16). They are in general somewhat different from those estimated with minimum contact distances as discussed by Ramakrishnan and Ramachandran (1965). However, whether they give better results than hard-sphere approximations remains to be seen.

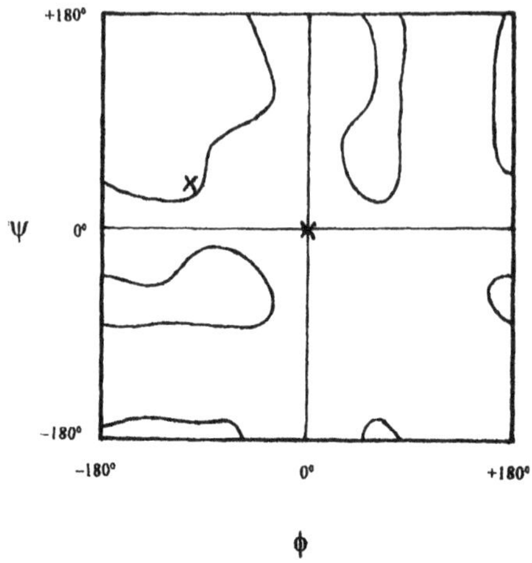

Figure 5-15. Allowed region in the (ϕ,ψ) plot for Ala based on interaction potentials.

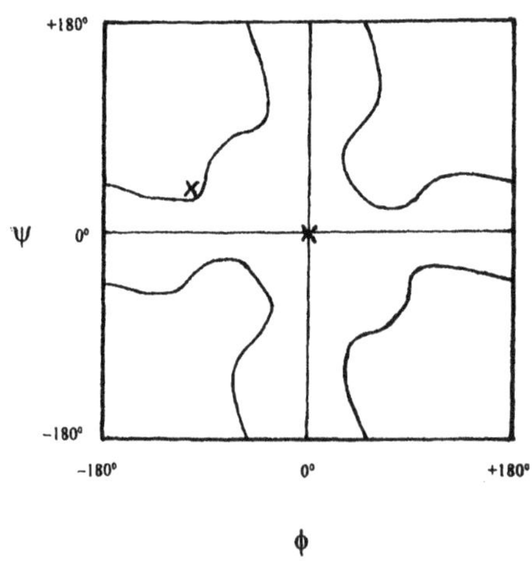

Figure 5-16. Allowed region in the (ϕ,ψ) plot for Gly based on interaction potentials.

DISTANCE BETWEEN NEAREST NEIGHBORING C_α–ATOMS

The distance between the two nearest neighboring C_α–atoms is independent of the (ϕ,ψ) angles as shown in Fig. 5-17.

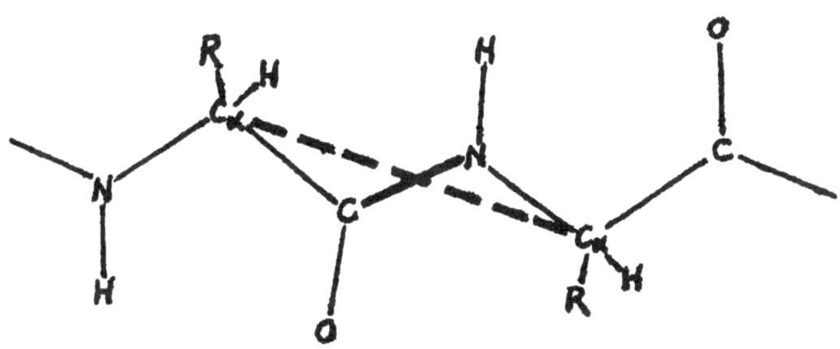

Figure 5-17. Distance between two nearest neighboring C_α–atoms.

By extending the line through C to N, we can calculate that distance as:

$$\{(0.62 + 1.32 + 0.80)^2 \ + \ (1.40 + 1.23)^2\}^{1/2}$$

$$= 3.80$$

Thus, it has a fixed distance of 3.80 A. Some of the old wire models of proteins were constructed with this fixed distance together with the angles between successive segments, known as the C_α–atom models.

SECONDARY STRUCTURES OF POLYPEPTIDES

If a linear polypeptide has the same (ϕ,ψ) angles for every amino acid residue, it will fold up into a periodic structure, commonly referred to as the secondary structure. The most well known one is the right-handed α–helix originally predicted by Pauling *et al.* (1951) theoretically, and verified experimentally soon afterwards. It consists of 3.6 amino acid residues per turn, and is stabilized by the presence of hydrogen-bonds between the backbone N-atoms and the carbonyl O-atoms. The ring structure thus formed contains 13 atoms. It is therefore sometimes referred to as a 3.6_{13} helix. In order to form these hydrogen-bonds, the (ϕ,ψ) angles are roughly in the range of ($-48^0,-57^0$) to ($-67^0,-44^0$) in the new convention, so that the α–helix can have some flexibility (Dickerson and Geis, 1969). In the case of myoglobin and hemoglobin (Chapter 2), nearly 80% of the protein is in the helical configuration. However, there is some debate as to whether these helices are α, i.e. 3.6_{13} , or 3_{10} which is slightly different with three amino acid residues per turn and the ring structure formed by hydrogen-bond having 10 atoms. The 3_{10}–helices have different (ϕ,ψ) angles (Dickerson and Geis, 1969).

Atomic models can be generated using these (ϕ,ψ) values, the translation and rotation of coordinates as discussed in the Appendix, and numerical computations with programming. These are very interesting studies but beyond the score of this book. However, using repeated peptide units represented in Fig. 5-3, it is possible to get an idea of the three dimensional structure of the α–helix.

Another important secondary structure is the β–strand. It is nearly straight with two amino acid residues per "turn". When aligned, several of them can form either anti-parallel or parallel β–pleated sheet with hydrogen-bonds. Several sheets may be stacked together for stability as in the case of antibody molecules (Chapter 1). For anti-parallel β–strands, their (ϕ,ψ) values are around (-139^0, $+135^0$); and for parallel β–strands, around ($-123^0,+116^0$). Similarly, these structures can be visualized by using the paper model illustrated in Fig. 5-3.

There are many other secondary or periodic structures of polypeptides, discussed by Dickerson and Geis (1969). Most of them are right-handed, but some are left-handed.

REFERENCES

Brant DA, Miller WG and Flory PJ (1967) Conformational energy estimates for statistically coiling polypeptide chains. *J. Mol. Biol.*, **23**, 47-65.

Clementi C, Vendruscolo M, Maritan A and Domany E (1999) Folding Lennard-Jones proteins by a contact potential. *Proteins.* **37**, 544-553.

Dickerson RE and Geis I (1969) *The Structure and Action of Proteins.* W. A. Benjamin, Inc., Menlo Park, CA.

Fogolari F, Esposito G, Viglino P and Cattarinussi S (1996) Modeling of polypeptide chains as C alpha chains, C alpha chains with C beta, and C alpha chains with ellipsoidal lateral chains. *Biophys. J.*, **70**, 1183-1197.

Pauling L , Corey RB and Branson HR (1951) *Proc. Nat. Acad. Sci. USA,* **37**, 207-

Ramakrishnan C and Ramachandran GN (1965) Stereochemical criteria for polypeptide and protein chain conformations II. Allowed configurations for a pair of peptide units. *Biophys. J.*, **5**, 909-933.

Schiff LI (1955) *Quantum Mechanics, 2nd Edition.* McGraw-Hill Book Company, Inc., New York.

White DN (1997) A computationally efficient alternative to the Buckingham potential for molecular mechanics calculations. *J. Comput. Aided Mol. Des.*, **11**, 517-521.

Wimberly BT, Brodersen DE, Clemons WM Jr, Morgan-Warren RJ, Carter AP, Vonrhein C, Hartsch T and Ramakrishnan V (2000) Structure of the 30S ribosomal subunit. *Nature,* **407**, 327-339.

EXERCISE

Calculate the bond distances, bond angles and (ϕ,ψ) angles from the following set of atomic coordinates:

	x	y	z
C_{-1}	16.683	21.745	12.868
N	17.852	21.693	12.260
C_α	18.733	20.517	12.317
C	19.008	20.156	13.779
N_{+1}	19.596	21.117	14.468

CHAPTER 6

PREDICTION OF PROTEIN FOLDING

INTRODUCTION

Since the classical experimental demonstration that the primary amino acid sequence of a protein contains the information for three-dimensional folding by Anfinsen (1973), numerous methods of prediction have been published in the literature. As we have discussed in the previous Chapter, proline and glycine residues can have important effects on the folding of the protein backbone.

Due to the absence of the H–atom on the backbone of the proline residue, it cannot form hydrogen-bonds in front of the N-terminal side of the protein chain. Thus, it is usually considered to be an α–helical breaking residue. In some cases, its presence results in a bend of the α–helix. At the same time, since its ϕ angle is around -60^0 which is somewhat different from that for either anti-parallel or parallel β–strand, proline is also considered as a β–strand breaker, or at least distorts a β–strand.

On the other hand, glycine residue can have many (ϕ,ψ) angle combinations, not possible for any other amino acid residue. Therefore, it has a tendency of being present in bends or turns of protein backbones.

As a result, many of the protein folding predictive methods rely on such statistical results of various amino acid residues being present in α–helices, β–strands, or in turns, as suggested from three-dimensional structures of proteins determined experimentally. Most of the methods are difficult to understand. One of these (Chou and Fasman, 1974) was relatively straight forward. However, sometimes one segment may have tendencies of being α–helical as well as forming β–strands. Rules were then proposed to choose one of these two possibilities. In general, most of the methods give good predictions for the presence of α–helices, not so good for β–strands, and even worse for turns.

The helical wheel method of predicting the existence of α–helix (Schiffer and Edmundsen, 1967) positions the amino acid residues in a protein along a theoretical α–helix. If there are many hydrophobic residues lining only one surface of that α–helix, it is assumed that the helix indeed exist (Fig. 6-1). The success of this method results from the fact that in most globular proteins, their α–helices are on the surface, with one side facing the interior of the protein, thus hydrophobic.

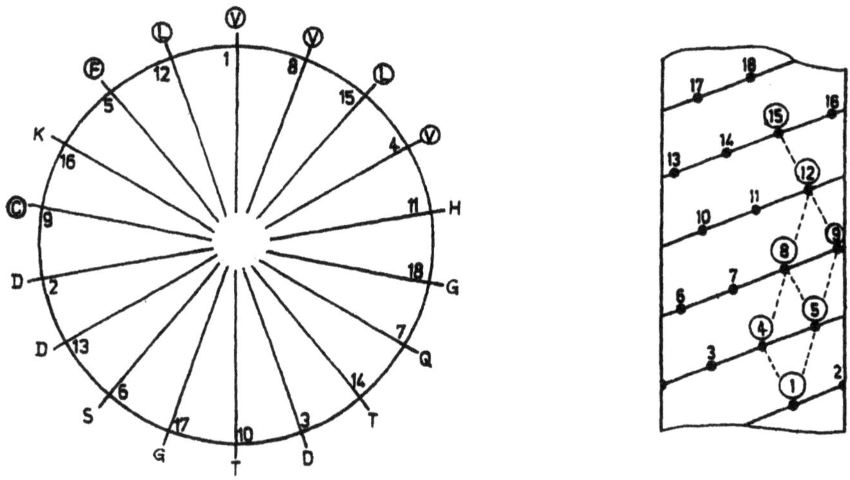

Figure 6-1. The helical wheel method.

The helical wheel method has also been used to predict trans-membrane regions of proteins. Due to the presence of a central hydrophobic layer in membranes, the α–helix should consist of hydrophobic amino acid residues. Since the rise per amino acid residue in an α–helix is 1.5 A, a stretch of around 18 hydrophobic residues in a protein strongly suggests the existence of a trans-membrane region of about 27 A in length.

For all these predictive methods, one may ask two questions:

(1) Is the prediction self-consistent?

(2) From the prediction, is it possible to display the three-dimensional structure of the protein?

Self-consistent means that if the three-dimensional structure of a protein is already determined experimentally, the predictive method should indeed predict that structure. The helical wheel method is self-consistent, but most of the other methods are not. If the locations of α–helices, β–strands, and turns are all precisely predicted, it is usually not possible to generate the three-dimensional structure of the protein. There are always amino acid residues not in one of these three configurations.

For these reasons, we will consider a predictive method involving the estimation of ϕ and ψ angles as defined in the previous Chapter. As discussed below, in principle, it is self-consistent and should be able to provide sufficient information to construct the entire three-dimensional structure.

REFERENCE PROTEINS

As we have discussed in the previous Chapter, there are many (ϕ,ψ) angles sterically allowed for all amino acid residues. Thus, for a protein of 100 amino acid residues, for example, if there are ten possible choices of (ϕ,ψ) angles for each residue, the total number of configurations to be analyzed will be 10^{100}. Even the fastest compute currently available will not be able to handle so many different calculations. It is thus essential to reduce the number of (ϕ,ψ) choices for every position of the protein.

Furthermore, experimentally observed (ϕ,ψ) angles as determined from atomic coordinates of proteins with known three-dimensional structures at the website, http://www.rcsb.org/pdb, may not agree with those predicted theoretically. Some of the experimentally determined (ϕ,ψ) angles are not within the allowed regions based on steric hindrance calculations. On the other hand, some of the allowed regions in the (ϕ,ψ) plot are not populated by experimentally observed values. In short, the simple-minded theoretical calculations discussed in the previous Chapter do not tell the entire story. Many factors are beyond our comprehension at this time.

Therefore, proteins with known three-dimensional structures will be used as references. Most of these are soluble proteins that have been crystallized. Even though their atoms cannot be visualized directly from X-ray diffraction studies due to insufficient resolution, atomic coordinates for all non-hydrogen atoms are usually provided with the incorporation of the standard peptide units. From these, the (ϕ,ψ) angles of each residues can be determined as discussed in detail in the precious Chapter. Together with the sequences of these reference proteins, their (ϕ,ψ) angles will provide the necessary information for predicting protein folding.

For example, the first 26 amino acid residues of myoglobin are listed in Table 6-1 together with their (ϕ,ψ) angles in the new convention. Due to the flexibility of the N-terminal end, the (ϕ,ψ) angles of Val at position 1 are not determined.

Table 6-1. (ϕ,ψ) angles of residues 2 to 26 of myoglobin.

Position	Amino Acid Residue	(ϕ,ψ) angles
1	Val	
2	Leu	$-47^0,+134^0$
3	Ser	$-50^0,+149^0$
4	Glu	$-48^0,-55^0$
5	Gly	$-54^0,-55^0$
6	Glu	$-53^0,-40^0$
7	Trp	$-62^0,-49^0$
8	Gln	$-40^0,-57^0$
9	Leu	$-64^0,-47^0$
10	Val	$-52^0,-49^0$
11	Leu	$-63^0,-41^0$
12	His	$-66^0,-35^0$
13	Val	$-66^0,-44^0$
14	Trp	$-51^0,-64^0$
15	Ala	$-41^0,-36^0$
16	Lys	$-73^0,-46^0$
17	Val	$-61^0,-32^0$
18	Glu	$-69^0,-9^0$
19	Ala	$-71^0,-35^0$

20	Asp	$-149^0,+69^0$
21	Val	$-48^0,-56^0$
22	Ala	$-61^0,-35^0$
23	Gly	$-61^0,-62^0$
24	His	$-49^0,-52^0$
25	Gly	$-55^0,-54^0$
26	Gln	$-48^0,-59^0$

Residues 4 to 17 all have (ϕ,ψ) angles close to the α–helical configuration. Residue 18 is more close to the 3_{10}–helical configuration (Dickerson and Geis, 1969). Residue 20 is definitely not α–helical. One may ask how accurate these values are. This turns out to be a question not usually answered by X-ray crystallographists. Instead, it is important to determine the three-dimensional structure by a different method.

For horse cytochrome c consisting of 104 amino acid residues, its three-dimensional structure has also been investigated by nuclear magnetic resonance (Kar *et al.,* 1994) measurements. There are about ten residues having quite different (ϕ,ψ) angles from those determined by X-ray diffraction studies. Such difference can be attributed to the different physical states, one being in solution and the other in crystal. However, they may also be due to the inability of positioning peptide units precisely into the electron density map provided by X-ray diffraction analysis. To connect the nearest C_α–atoms with backbone atoms, it is possible to offset the ψ angle by 180^0 and the ϕ angle of the following residue also by 180^0. Hopefully, such errors are relatively rare.

Thus, for all proteins with known atomic coordinates of their non-hydrogen atoms, the (ϕ,ψ) angles of each amino acid residue can be calculated, except those at the N-terminal and C-terminal ends. They are associated with the known amino acid sequences of these proteins. If indeed the sequence determines the three-dimensional folding of a protein, these (ϕ,ψ) angles can be closely correlated with the sequence. This valuable experimental information unfortunately cannot be derived theoretically, without extremely complicated calculations.

SELECTING ϕ AND ψ ANGLES

In order to predict the three-dimensional folding of any protein with known amino acid sequence, it is necessary to compare segments of its sequence with those in reference proteins. If matched, there is a possibility that locally the two identical segments may fold similarly. However, the chance of finding such matches is very small. On random basis, for a segment consisting of m amino acid residues, the chance is 20^{-m}. Thus, to estimate the (ϕ,ψ) angles of a specific amino acid residue due to the influence of neighboring ones on both sides, we have:

m = 1: 1 out of 20,

m = 3: 1 out of 8,000,

m = 5: 1 out of 3,200,000, etc.

It becomes clear that this matching process may be successful for m = 3, with the presently available reference proteins. Hopefully, as more three-dimensional structures of proteins are determined, m = 5 may be use. These values of m can barely include the effect of neighboring amino acid residues to promote the formation of α–helices.

On the other hand, β–pleated sheets are stabilized by hydrogen-bonds between different β–strands which can be positioned at various parts of the amino acid sequence. This long-range effect will be very difficult to predict using the method of matching sequences locally.

Various rules of selection can be designed to estimate the (ϕ,ψ) choices for the middle amino acid residue in tri-peptides (m = 3). An example is illustrated here. Consider the tri-peptide, Val Glu Lys (Wu *et al.*,1974). This sequence can be matched exactly with tri-peptides in reference proteins. One of the (ϕ,ψ) angles found in these reference proteins is $(-75^{\circ},+142^{\circ})$. On the other hand, out of the 8,000 possible tri-peptides, a fairly large number of them are not present in reference proteins. Thus, in addition, for every tri-peptide, its constituent di-peptides will also be matched. In the present case, some of the (ϕ,ψ) angles found in reference proteins for Val Glu ( ), Val ( ) Lys, and ( ) Glu Lys, are listed in Table 6-2.

Table 6-2. Some of the (ϕ,ψ) angles found in reference proteins for the three di-peptides.

Val Glu ( )	Val ( ) Lys	( ) Glu Lys
$-69^0,-9^0$	$-101^0,-154^0$	$-46^0,-34^0$
$-82^0,-33^0$	$-128^0,+136^0$	$-48^0,-43^0$
$-45^0,-42^0$	$-133^0,+147^0$	$-114^0,+180^0$
$-121^0,+134^0$	$-104^0,+125^0$	$-124^0,-30^0$
$-65^0,-42^0$	$-61^0,-49^0$	$-99^0,-29^0$
$-29^0,-45^0$	$-67^0,-59^0$	$-87^0,-44^0$
$-68^0,-56^0$	$-89^0,+92^0$	$-93^0,+22^0$
$-60^0,-39^0$		

If the ϕ and ψ values for all three di-peptides in Table 6-2 are within about 30^0 of each other, they are combined and averaged. They give a possible choice of $(-58^0,-41^0)$. Other less restrictive selections are also possible.

This procedure of matching amino acid sequences of tri-peptides and selecting the (ϕ,ψ) angles of the middle amino acid residue of reference proteins is self-consistent. Consider a protein with known three-dimensional structure. If this protein sequence together with its (ϕ,ψ) angles is included in the reference proteins, the process of matching tri-peptides will automatically pick the known (ϕ,ψ) angles as possible choices. This seemingly trivial result, however, provides a consistent scheme of selecting ϕ and ψ angles. A known three-dimensional structure will automatically be included as one of the possible choices.

Consider the above mentioned case of Val Glu Lys with two possible (ϕ,ψ) choices for the middle amino acid residue Val, i.e. $(-75^0,+142^0)$ and $(-59^0,-41^0)$. If every residue in a protein with 100 amino acid residues has two possible choices of (ϕ,ψ) angles, a total of 2^{100} or roughly 10^{30} possible combinations have to be analyzed. This number, although much smaller than 10^{100}, is still too large to be processed by modern day computers. Even though the (ϕ,ψ) angles of every residue together with the structure of the standard peptide unit can in principle generate detailed three-dimensional structures of the protein, it would be impossible to pick the correct one. In addition, the variation of as small as 1^0 for the selected ϕ and ψ angles can produce extensive distortions on cumulative basis over 100 residues.

BACKBONE STRUCTURE PREDICTION

Therefore, further reduction of (ϕ,ψ) choices for each amino acid residue seems to be essential for this matching process to be useful in predicting three-dimensional structures of proteins. Essentially only one choice for each residue will be needed. One way of achieving this goal is to consider all other proteins homologous to the one under study. For example, presumably cytochromes c from various species should fold into similar or nearly identical three-dimensional structures. However, locally, their amino acid sequences may be different. For a certain position, the (ϕ,ψ) choices can be different for each of these sequences. The required choice for the collection of all these sequences must be common to all of them. As a result, some of the positions of cytochromes c may have only one choice for their (ϕ,ψ).angles. Other positions can still have several choices.

Short Peptide Hormones

Another possible application of this matching process is to short peptides such as the nine amino acid neurohypophyseal hormones (Honig *et al.*, 1973). Even in the case of several (ϕ,ψ) choices for each amino acid residue, the total number of combination of choices is relatively small since there are only a few residues. These peptide hormones do not form crystals easily. Thus their three-dimensional structures cannot be analyzed by X-ray diffraction studies. On the other hand, nuclear magnetic resonance (NMR) can be a powerful physical tool to determine their structures. Thus, results based on theoretical predictive methods can be compared with experimental NMR findings, although the solvent in the NMR study may not be water.

Homologous sequences are also used to reduce the number of (ϕ,ψ) choices. They are listed in Table 6-3. The presence of a disulfide bond between Cys 1 and Cys 6 can provide additional restrictions of the (ϕ,ψ) angles. However, the N-terminal and C-terminal residues, i.e. Cys 1 and Gly 9, can have some flexibility. Energy minimization with various potential functions between non-bounded atoms as discussed in the previous Chapter can be used to refine the (ϕ,ψ) angles. In the process, the Cys 1 and Cys 6 disulfide bond is established and steric hindrance avoided.

Table 6-3. Amino acid sequences of seven neurohypophyseal hormones. All C-terminal Gly are amidated. Cys 1 and Cys 6 form a disulfide bond.

	1	2	3	4	5	6	7	8	9
Arginine-vasopressin	Cys	Tyr	Phe	Gln	Asn	Cys	Pro	Arg	Gly
Lysine-vasopressin	Cys	Tyr	Phe	Gln	Asn	Cys	Pro	Lys	Gly
Vasotocin	Cys	Tyr	Ile	Gln	Asn	Cys	Pro	Arg	Gly
Oxytocin	Cys	Tyr	Ile	Gln	Asn	Cys	Pro	Leu	Gly
Mesotocin	Cys	Tyr	Ile	Gln	Asn	Cys	Pro	Ile	Gly
Isotocin	Cys	Tyr	Ile	Ser	Asn	Cys	Pro	Ile	Gly
Glumitocin	Cys	Tyr	Ile	Ser	Asn	Cys	Pro	Gln	Gly

For these seven neurohypophyseal hormones listed in Table 6-3, the initial choices of (ϕ,ψ) angles for the nine positions are given in Table 6-4, together with various additional adjustments.

Table 6-4. Initial choices and adjustments of the (ϕ,ψ) angles for neurohypophyseal hormones.

Position	Initial choices	Minimizing Cys1-Cys 6 distance	Energy minimization
1	$-135^0,+166^0$	$-135^0,+166^0$	$-152^0,+140^0$
2	$-60^0,-60^0$; $-100^0,+130^0$	$-60^0,-60^0$	$-80^0,-60^0$
3	$-66^0,-11^0$	$-66^0,-11^0$	$-67^0,-36^0$
4	$-78^0,+142^0$	$-78^0,+142^0$	$-98^0,+149^0$
5	$-96^0,+32^0$; $-65^0,-25^0$	$-96^0,+32^0$	$-99^0,+34^0$
6	$-135^0,+154^0$; $-52^0,+141^0$	$-52^0,+141^0$	$-43^0,+147^0$
7	$-53^0,-38^0$	$-53^0,-38^0$	$-53^0,-21^0$
8	$-59^0,-46^0$; $-97^0,+139^0$	$-59^0,-46^0$	$-72^0,-13^0$
9	$+81^0,-30^0$	$+81^0,-30^0$	$+82^0,+69^0$

Other models may be constructed with the additional requirement of binding Cu^{++} (Honig *et al.*, 1973).

Peptide Loops on Proteins

Typical examples of peptide loops on proteins are the complementarity determining regions (CDR's) of antibodies discussed in Chapter 1. Numerous amino acid sequences of CDR's have been determined during the past 35 years. The method of predicting peptide backbone folding has been applied to these sequences, and results in theoretically generating detailed three-dimensional structures of antibody combining sites. An examples will be illustrated here.

In this case, the three-dimensional structure of the framework regions of a similar antibody is used as a scaffold. Structures of various CDR's are predicted by matching sequences and (ϕ,ψ) angles to reference proteins, and connected onto the scaffold. Therefore, the predicted loop folding must have the N-terminal residue starting at a fixed position in space with a given orientation. At the end of the loop, the C-terminal residue also has a fixed location and orientation. Furthermore, steric hindrance should be avoided not only within the same CDR loop but also among different CDR loops. These restrictions are usually sufficient to reduce the possible theoretical three-dimensional structures of the six CDR loops to just one or at most a few choices.

This approach has been applied to immunoglobulin MOPC-315 (Stanford and Wu, 1981). The known atomic coordinates of the framework regions for another immunoglobulin Newm (Poljak *et al.*,1972; and the website, http://www.rcsb.org/pdb) were used as the scaffold.

For each CDR, the known (ϕ,ψ) angles of the amino acid residue immediately to the N-terminal side is included to fix the location and orientation of that end. In order to fix the location and orientation of the C-terminal end, the known (ϕ,ψ) angles of the three amino acid residues immediately after the CDR segment are also included. The (ϕ,ψ) choices of the CDR amino acid residues are then adjusted to make sure that each CDR loop can fit onto the scaffold. The criterion of fitting again depends on minimizing a sum of squares. In this case, they are the distances between experimental and theoretical locations of the C_α–atoms of the three amino acid residues immediately after the C-terminal end of the CDR. By adjusting the (ϕ,ψ) angles of the CDR amino acid residues and adjacent ones, this sum of squares can usually be reduced to less than 1.5 A^2. However, it is important to avoid steric hindrance. Since Newm has no CDRL2, a longer segment from MOPC-315 is used.

For the six CDR's, the structures of the middle ones, i.e. CDRH3 and CDRL3, are predicted first, followed by those of CDRH1 and CDRL1. CDRH2 and CDRL2 are on the periphery and their structures are usually predicted last. The six CDR's form a very compact structure with little flexibility. In the case of MOPC-315, the final sets of (ϕ,ψ) angles for the six CDR's are illustrated in Tables 6-5. Angles are rounded off to the nearest degree. Table 6-6 gives the predicted atomic coordinates of the non-hydrogen backbone atoms. These results were from Stanford (1979).

Table 6-5. Amino acid residue (ϕ,ψ) angles of MOPC-315 CDR's. Adjacent residues are in parenthesis.

Position (Kabat numbering)		Amino acid residue	(ϕ,ψ) angle
CDRL1	(23)	Cys	$-130^0,+133^0$
	24	Arg	$-155^0,+131^0$
	25	Ser	$-72^0,-41^0$
	26	Ser	$-75^0,+151^0$
	27	Thr	$-112^0,-16^0$
	27A	Gly	$-79^0,-41^0$
	27B	Ala	$-107^0,+154^0$
	27C	Val	$-63^0,-47^0$
	28	Thr	$-138^0,+139^0$
	29	Thr	$-118^0,+137^0$
	30	Ser	$+53^0,+82^0$
	31	Asn	$-66^0,+108^0$
	32	Tyr	$-157^0,+151^0$
	33	Ala	$-69^0,+99^0$
	34	Asn	$-124^0,+91^0$
	(35)	Trp	$-142^0,+140^0$
	(36)	Ile	$-147^0,+123^0$
	(37)	Gln	$-130^0,+123^0$
CDRL2	(46)	Gly	$-86^0,+106^0$
	(47)	Leu	$-72^0,-37^0$
	(48)	Ile	$-152^0,+133^0$
	(49)	Gly	$-125^0,+149^0$
	50	Gly	$-67^0,-33^0$
	51	Thr	$-79^0,-38^0$
	52	Ser	$-144^0,+111^0$

	53	Asp	−100°,−31°
	54	Arg	−76°,+69°
	55	Ala	−75°,+122°
	56	Pro	−64°,+142°
	(57)	Gly	+65°,+15°
	(58)	Val	−93°,+137°
	(59)	Pro	−77°,+166°
	(60)	Val	−76°,−8°
	(61)	Arg	−57°,−17°
	(62)	Phe	−108°,+147°
	(63)	Ser	−168°,+140°
	(64)	Gly	−104°,+151°
CDRL3	(88)	Cys	−121°,+118°
	89	Ala	−103°,+125°
	90	Leu	−168°,−178°
	91	Trp	−125°,+132°
	92	Phe	−120°,−179°
	93	Arg	−71°,−42°
	94	Asn	−61°,−72°
	95	His	−161°,+133°
	96	Phe	−58°,+2°
	97	Val	−41°,+147°
	(98)	Phe	−89°,+116°
	(99)	Gly	−101°,+97°
	(100)	Gly	−100°,+141°
CDRH1	(30)	Thr	−45°,−76°
	31	Ser	+38°,−112°
	32	Gly	−122°,+146°
	33	Tyr	−109°,+152°
	34	Phe	−177°,+176°
	35	Trp	−57°,+119°
	35A	Asn	−110°,+34°
	(36)	Trp	−136°,+91°
	(37)	Ile	−109°,+132°
	(38)	Arg	−120°,+164°
CDRH2	(47)	Trp	−74°,+137°
	(48)	Leu	−46°,−73°
	(49)	Gly	−170°,−158°
	50	Phe	−112°,+141°
	51	Ile	−95°,+150°
	52	Lys	−134°,+137°
	53	Tyr	−66°,−38°

	54	Asp	$-107^0,-9^0$
	55	Gly	$-35^0,+160^0$
	56	Ser	$-81^0,+128^0$
	57	Asx	$-113^0,-3^0$
	58	Tyr	$-71^0,+1^0$
	59	Gly	$-96^0,+142^0$
	60	Asx	$-90^0,+162^0$
	61	Pro	$-77^0,+122^0$
	62	Ser	$-159^0,+165^0$
	63	Leu	$-72^0,-22^0$
	64	Lys	$-69^0,+159^0$
	65	Asn	$-46^0,-22^0$
	(66)	Arg	$-70^0,+64^0$
	(67)	Val	$-109^0,+93^0$
	(68)	Ser	$-106^0,+134^0$
CDRH3	(94)	Gly	$-114^0,+154^0$
	95	Asp	$-116^0,+28^0$
	96	Asn	$-53^0,+138^0$
	97	Asp	$-61^0,-56^0$
	98	His	$-165^0,+88^0$
	99	Leu	$-143^0,+162^0$
	100	Tyr	$-83^0,+32^0$
	100A	Phe	$-90^0,+38^0$
	101	Asp	$-62^0,+133^0$
	102	Tyr	$-72^0,-70^0$
	(103)	Trp	$-118^0,+128^0$
	(104)	Gly	$-123^0,+97^0$
	(105)	Gln	$+17^0,+148^0$

Table 6-6. Atomic coordinates of non-hydrogen backbone atoms of MOPC-315 CDR's.

Non-hydrogen atom		x	y	z
CDRL1	24 N	-38.29	15.75	-6.62
	24 CA	-39.78	16.00	-7.28
	24 C	-40.64	15,05	-6.72
	25 N	-41.76	15,65	-6.36
	25 CA	-42.91	14.92	-5.79
	25 C	-43.58	14.10	-6.89
	26 N	-43.67	14.72	-8.05
	26 CA	-44.29	14.10	-9.24
	26 C	-43.33	13.07	-9.83

	27 N	-43.93	12.08	-10.46
	27 CA	-43.20	10.98	-11.10
	27 C	-43.37	11.07	-12.62
	27A N	-44.35	11.86	-13.01
	27A CA	-44.69	12.09	-14.42
	27A C	-43.71	13.10	-15.02
	27B N	-43.42	14.11	-14.21
	27B CA	-42.50	15.19	-14.58
	27B C	-41.19	15.03	-13.82
	27C N	-40.15	15.57	-14.43
	27C CA	-38.79	15.53	-13.87
	27C C	-38.76	16.30	-12.55
	28 N	-39.35	17.48	-12.60
	28 CA	-39.43	18.39	-11.44
	28 C	-40.83	19.01	-11.39
	29 N	-41.33	19.09	-10.17
	29 CA	-42.66	19.65	-9.88
	29 C	-42.51	20.88	-8.98
	30 N	-43.27	21.90	-9.34
	30 CA	-43.29	23.18	-8.62
	30 C	-41.86	23.71	-8.49
	31 N	-41.42	24.35	-9.57
	31 CA	-40.08	24.93	-9.66
	31 C	-39.96	26.09	-8.67
	32 N	-39.19	25.82	-7.63
	32 CA	-38.93	26.79	-6.55
	32 C	-37.62	26.44	-5.85
	33 N	-37.00	27.48	-5.33
	33 CA	-35.72	27.36	-4.61
	33 C	-35.95	26.64	-3.28
	34 N	-35.64	25.36	-3.31
	34 CA	-35.78	24.47	-2.14
	34 C	-34.43	23.81	-1.83
CDRL2	50 N	-36.44	30.53	-1.17
	50 CA	-37.42	30.98	-2.16
	50 C	-37.75	32.46	-1.92

	51 N	-37.72	32.81	-0.64
	51 CA	-38.00	34.18	-0.19
	51 C	-36.77	35.06	-0.41
	52 N	-35.62	34.45	-0.18
	52 CA	-34.32	35.11	-0.34
	52 C	-33.30	34.10	-0.88
	53 N	-32.91	34.34	-2.11
	53 CA	-31.94	33.50	-2.82
	53 C	-30.56	34.16	-2.78
	54 N	-30.60	35.48	-2.75
	54 CA	-29.39	36.32	-2.71
	54 C	-28.80	36.30	-1.30
	55 N	-28.29	35.13	-0.95
	55 CA	-27.67	34.89	0.36
	55 C	-26.29	35.55	0.41
	56 N	-26.15	36.43	1.38
	56 CA	-24.91	37.19	1.61
	56 C	-23.79	36.22	1.99
CDRL3	89 N	-36.62	22.30	0.93
	89 CA	-37.85	22.73	1.61
	89 C	-38.74	23.50	0.63
	90 N	-39.96	23.02	0.52
	90 CA	-40.98	23.61	-0.36
	90 C	-42.35	23.01	-0.03
	91 N	-43.32	23.45	-0.80
	91 CA	-44.72	23.01	-0.66
	91 C	-45.22	22.46	-2.00
	92 N	-45.83	21.29	-1.90
	92 CA	-46.39	20.59	-3.06
	92 C	-47.90	20.39	-2.87
	93 N	-48.49	19.75	-3.85
	93 CA	-49.93	19.45	-3.87
	93 C	-50.24	18.38	-2.82
	94 N	-49.36	17.40	-2.78
	94 CA	-49.47	16.26	-1.85
	94 C	-49.42	16,78	-0.41

	95 N	-48,23	17.20	-0.03
	95 CA	-47.96	17.74	1.31
	95 C	-46.67	18,55	1.29
	96 N	-46.76	19.73	1.88
	96 CA	-45.64	20.67	1.97
	96 C	-44.46	20.00	2.68
	97 N	-44.69	18.74	3.01
	97 CA	-43.69	17.91	3.70
	97 C	-42.30	18.17	3.11
CDRH1	31 N	-44.22	39.79	12.07
	31 CA	-43.79	39.50	11.70
	31 C	-44.74	38.48	11.07
	32 N	-44.19	37.30	10.84
	32 CA	-44.92	36.17	10.24
	32 C	-44.88	34.98	11.19
	33 N	-45.97	34.24	11.17
	33 CA	-46.15	33.04	12.00
	33 C	-46.09	31.79	11.12
	34 N	-45.66	30.71	11.75
	34 CA	-45.53	29.40	11.09
	34 C	-45.07	28.36	12.11
	35 N	-45.01	27.13	11.64
	35 CA	-44.59	25.98	12.45
	35 C	-43.20	26.23	13.03
	35A N	-43.17	26.45	14.35
	35A CA	-41.94	26.48	15.12
	35A C	-41.52	25.18	15.81
CDRH2	50 N	-48.94	21.81	10.53
	50 CA	-49.40	23.15	10.16
	50 C	-50.17	23.77	11.33
	51 N	-49.94	25.06	11.50
	51 CA	-50.57	25.85	12.57
	51 C	-51.81	26.54	12.02
	52 N	-52.76	26.74	12.92
	52 CA	-54.04	27.39	12.61
	52 C	-54.36	28.43	13.68

53 N	-54.93	29.57	13.21
53 CA	-55.20	30.70	14.07
53 C	-56.41	30.31	14.92
54 N	-57.19	29.40	14.37
54 CA	-58.41	28.89	15.02
54 C	-58.16	27.46	15.50
55 N	-57.02	26.94	15.09
55 CA	-56.59	25.58	15.44
55 C	-57.05	25.25	16.86
56 N	-57.09	23.96	17.13
56 CA	-57.50	23.42	18.43
56 C	-56.52	23.50	19.41
57 N	-56.61	24.09	20.55
57 CA	-55.62	24.27	21.63
57 C	-56.03	23.42	22.83
58 N	-57.17	22.77	22.68
58 CA	-57.75	21.90	23.71
58 C	-56.91	20.63	23.83
59 N	-55.87	20.59	23.01
59 CA	-54.94	19.46	22.96
59 C	-53.72	19.76	23.84
60 N	-53.27	18.71	24.49
60 CA	-52.11	18.77	25.40
60 C	-50.83	18.48	24.61
61 N	-49.72	18.87	25.21
61 CA	-48.38	18.70	24.63
61 C	-47.95	17.24	24.79
62 N	-47.67	16.63	23.65
62 CA	-47.23	15.23	23.58
62 C	-46.50	15.00	22.25
63 N	-45.90	13.88	22.21
63 CA	-45.23	13.46	21.04
63 C	-45.98	13.01	19.93
64 N	-47.18	12.66	20.36
64 CA	-48.25	12.20	19.47
64 C	-48.74	13.37	18.61
65 N	-49.36	13.00	17.50
65 CA	-49.90	13.96	16.64
65 C	-50.68	15.05	17.28

CDRH3	95 N	-39.74	31.73	10.01
	95 CA	-39.48	33.82	8.42
	95 C	-39.35	35.17	6.39
	96 N	-39.48	33.82	8.42
	96 CA	-39.34	35.19	7.91
	96 C	-39.35	35.17	6.39
	97 N	-40.06	36.14	5.84
	97 CA	-40.21	36.31	4.39
	97 C	-38.83	36.56	3.77
	98 N	-38.18	37.58	4.28
	98 CA	-36.84	37.99	3.83
	98 C	-36.23	38.94	4.86
	99 N	-35.56	38.34	5.82
	99 CA	-34.89	39.06	6.91
	99 C	-33.59	38.36	7.27
	100 N	-32.74	39.10	7.95
	100 CA	-31.42	38.62	8.39
	100 C	-31.58	37.84	9.70
	100A N	-32.72	37.20	9.80
	100A CA	-33.08	36.38	10.97
	100A C	-32.61	34.94	10.75
	101 N	-32.06	34.39	11.82
	101 CA	-31.55	33.01	11.83
	101 C	-32.69	32.04	11.66
	102 N	-32.40	31.12	10.65
	102 CA	-33.35	30.08	10.23
	102 C	-33.52	29.06	11.35

Comparison with Experimental Results

As discussed in Chapter 1, CDRH3 of an antibody molecule confers fine specificity. Their structures have been determined experimentally in several antibody crystal structures (Kabat *et al.*, 1991). Due to the "anchoring" amino acid residues at the two ends of CDRH3, they all form loops (Figs. 6-2 to 6-11), similar to the above prediction.

Figure 6-2. Stereo view of CDRH3 with five amino acid residues.

Figure 6-3. Stereo view of CDRH3 with six amino acid residues.

Figure 6-4. Stereo view of CDRH3 with seven amino acid residues.

Figure 6-5. Stereo view of CDRH3 with eight amino acid residues.

Figure 6-6. Stereo view of CDRH3 with nine amino acid residues.

Figure 6-7. Stereo view of CDRH3 with ten amino acid residues.

Figure 6-8. Stereo view of CDRH3 with eleven amino acid residues.

Figure 6-9. Stereo view of CDRH3 with twelve amino acid residues.

Figure 6-10. Stereo view of CDRH3 with fifteen amino acid residues.

3D6

Figure 6-11. Stereo view of CDRH3 with seventeen amino acid residues.

As a result, depending on the number of amino acid residues, some canonical loops (Chothia and Lesk, 1987) have been proposed without the detailed analysis of selecting ϕ and ψ angles. Thus, for antibodies, an abbreviated approach of predicting backbone structures has been proposed with the use of such standard configurations for all six CDR's (Webster and Rees, 1995).

Proteins of Similar Functions

Many proteins have similar enzymatic functions, even though their amino acid sequences are different. For example, as discussed in Chapter 4, glycerol dehydrogenase uses NAD as a co-enzyme. There are many other NAD-linked dehydrogenases and have similar three-dimensional structures consisting of parallel β–pleated sheets stabilized by α–helices. Thus, it is possible to make use of a known dehydrogenase structure as a framework to predict the three-dimensional folding of another one.

Several of the NAD-linked dehydrogenases have an NAD domain consisting of a β–pleated sheet with six parallel β–strands and a left-handed twist of about 100°. That sheet is stabilized by four α–helices, two on each side of the sheet. However, there is little amino acid sequence homology among these dehydrogenases, except that Gly residues are common at ends of several β–strands and an Asp residue is conserved for NAD binding. Although the catalytic domains of these dehydrogenases appear somewhat different, there are still many structural similarities.

With these restrictions and other information derived from mutational studies, symmetry of tetramer subunits, etc., the three-dimensional structure of the backbone of ribitol dehydrogenase (Stevens and Wu, 1985) has been proposed based on the predictive method discussed in this Chapter. While it is not sufficient to select (ϕ,ψ) angles for each amino acid residue of a protein to predict its three-dimensional folding, the incorporation of additional information can, in some cases, provide sufficient restrictions to predict the tertiary structure of the backbone of a protein.

With the rapid development of various genome projects, the amino acid sequences of many proteins, some represented by the so-called open reading frames (ORF's), will become available. Prediction of their three-dimensional folding and possible structures can provide valuable biological information to their functions.

SIDE-CHAIN STRUCTURE PREDICTION

After the backbone structure of a peptide loop, e.g. one of the CDR's of antibodies, is proposed, the side chains of all amino acid residues in the loop can be analyzed. An example is illustrated for an immunoglobulin, MOPC104E (McIntire *et al.*, 1965). It is an IgM with a λ light chain Appella *et al.*, 1968) and a μ heavy chain (Kehry *et al.*, 1979). It reacts with dextran, especially dextran B1355 from *L. mesenteroides*, thus specific for the $\alpha(1\rightarrow3)$ linkage. Nigertriose can very effectively inhibit MOPC104E's binding of dextran B1355, suggesting that the binding site may consist of three residues of $\alpha(1\rightarrow3)$-linked glucose. Further study (Schepers *et al.*, 1978) suggests that there is a 12 A-deep cavity complementary to the nigerose disaccharide with a subsite for a third glucose residue in an adjacent groove. MOPC104E is thus specific for terminal nigerosyl residues as in the case of dextran B1355, but not for $\alpha(1\rightarrow3)$ linkages only in the middle of dextran molecules.

Various attempts to crystallize MOPC104E for X-ray diffraction studies have unfortunately failed. Thus, its detailed antibody combining site three-dimensional structure can only be predicted (Hovis and Wu, 1985). Similar to MOPC315, the predicted CDR's are joined onto the framework structure of Newm, except in the region of CDRL2 where that of REI (Epp *et al.*, 1975) is also used. In the order of prediction, the CDR amino acid sequences and their (ϕ,ψ) angles are listed in Tables 6-7 to 6-12 (Hovis, 1982).

Table 6-7. Amino acid sequence and (ϕ,ψ) angles of CDRL3 of MOPC104E.

Position (Kabat numbering)	Amino acid residue	(ϕ,ψ) angles
89	Ala	$-92^0,-35^0$
90	Leu	$-81^0,-9^0$
91	Trp	$-42^0,-38^0$
92	Tyr	$-115^0,+139^0$
93	Ser	$-56^0,-11^0$
94	Asn	$-110^0,+118^0$
95	His	$-61^0,-38^0$
96	Trp	$-63^0,-21^0$
97	Val	$-100^0,+145^0$

Table 6-8. Amino acid sequence and (ϕ,ψ) angles of CDRH3 of MOPC104E.

Position (Kabat numbering)	Amino acid residue	(ϕ,ψ) angles
95	Asp	$-53^0,-52^0$
96	Tyr	$-114^0,+137^0$
97	Asp	$-92^0, +79^0$
98	Trp	$-69^0,-11^0$
99	Tyr	$-127^0,+161^0$
100	Phe	$-44^0,-48^0$
101	Asp	$-114^0,+128^0$
102	Val	$-110^0,-4^0$

Table 6-9. Amino acid sequence and (ϕ,ψ) angles of CDRH1 of MOPC104E.

Position (Kabat numbering)	Amino acid residue	(ϕ,ψ) angles
31	Asp	$-32^0,-74^0$
32	Tyr	$-105^0,+128^0$
33	Tyr	$-123^0,+180^0$
34	Met	$-82^0,+158^0$
35	Lys	$-135^0,+177^0$

Table 6-10. Amino acid sequence and (ϕ,ψ) angles of CDRL1 of MOPC104E.

Position (Kabat numbering)	Amino acid residue	(ϕ,ψ) angles
24	Arg	$-129^0,+161^0$
25	Ser	$-127^0,+170^0$
26	Ser	$-50^0,-37^0$
27	Thr	$-149^0,+163^0$
27A	Gly	$-12^0,+94^0$
27B	Ala	$-60^0,-52^0$
27C	Val	$-48^0,+131^0$
28	Thr	$-139^0,+134^0$
29	Thr	$-81^0,-6^0$
30	Ser	$-60^0,-44^0$
31	Asn	$-120^0,+142^0$
32	Tyr	$-92^0,-13^0$
33	Ala	$-72^0,+101^0$
34	Asn	$-71^0,-67^0$

Table 6-11. Amino acid sequence and (ϕ,ψ) angles of CDRL2 of MOPC104E.

Position (Kabat numbering)	Amino acid residue	(ϕ,ψ) angles
50	Gly	$-65^0,-55^0$
51	Thr	$-62^0,-46^0$
52	Asn	$-64^0,+138^0$
53	Asn	$-72^0,-44^0$
54	Arg	$-168^0,+172^0$
55	Ala	$-109^0,+138^0$
56	Pro	$-73^0,+136^0$

Table 6-12. Amino acid sequence and (ϕ,ψ) angles of CDRH2 of MOPC104E.

Position (Kabat numbering)	Amino acid residue	(ϕ,ψ) angles
50	Asp	$-177^0,+154^0$
51	Ile	$-129^0,+127^0$
52	Asn	$-138^0,+120^0$
52A	Pro	$-69^0,+7^0$
53	Asn	$-62^0,-61^0$
54	Asn	$-81^0,-41^0$

55	Gly	+133°,−25°
56	Gly	+67°,+36°
57	Thr	−64°,−10°
58	Ser	−153°,+117°
59	Tyr	−67°,+124°
60	Asn	−84°,+167°
61	Gln	−136°,+125°
62	Lys	−68°,−37°
63	Phe	−40°,−37°
64	Lys	+23°,+76°
65	Gly	+162°,−79°

Based on these backbone structures of the MOPC104E CDR's, the side chains of each amino acid residue can then be adjusted to avoid steric hindrance of non-bounded atoms. Their three-dimensional structures are shown in Figs. 6-12 to 6-17.

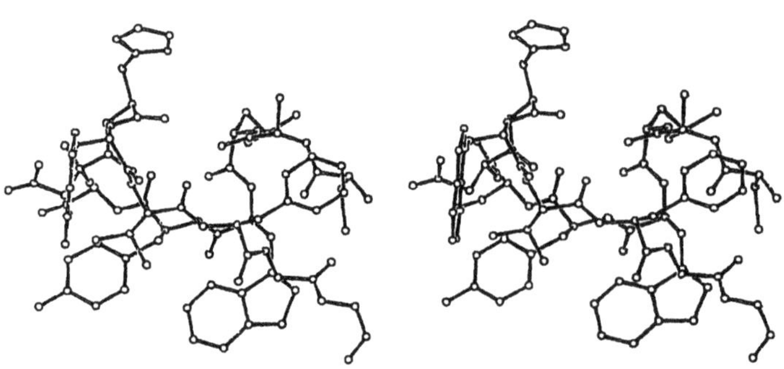

Figure 6-12. Stereo view of CDRL3 of MOPC104E.

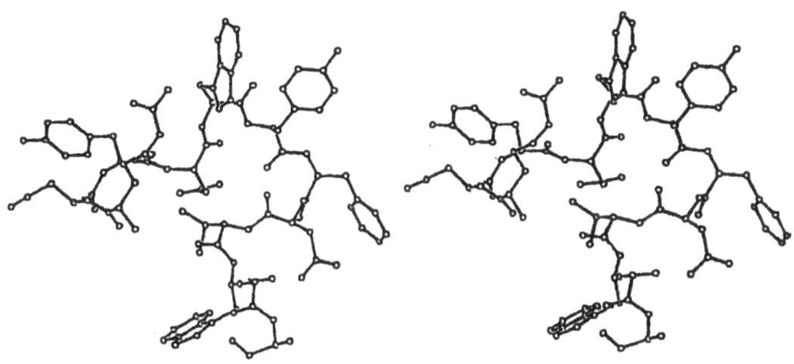

Figure 6-13. Stereo view of CDRH3 of MOPC104E.

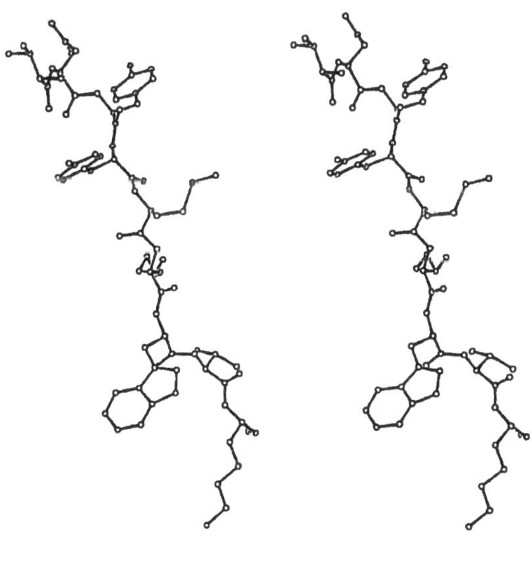

Figure 6-14. Stereo view of CDRH1 of MOPC104E.

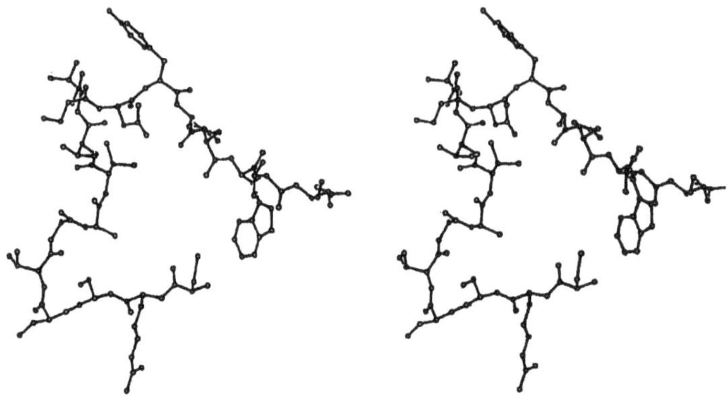

Figure 6-15. Stereo view of CDRL1 of MOPC104E.

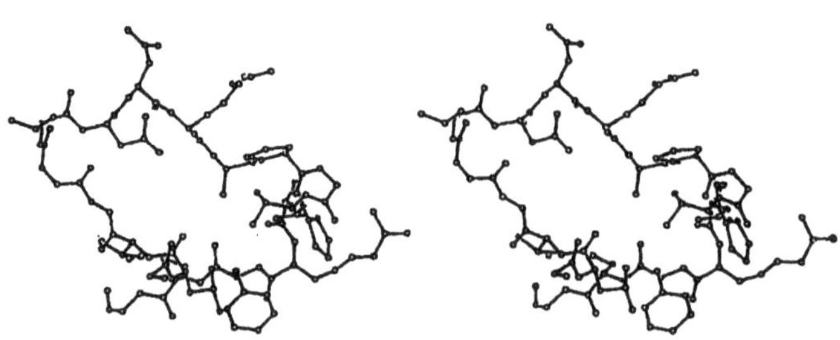

Figure 6-16. Stereo view of CDRL2 of MOPC104E.

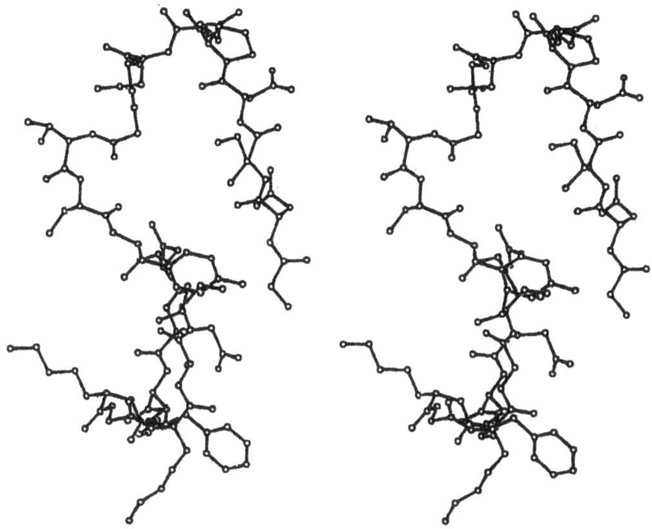

Figure 6-17. Stereo view of CDRH2 of MOPC104E.

Indeed, the atomic coordinates of all non-hydrogen atoms are determined (Hovis, 1982).

FUTURE THEORETICAL STRUCTURAL STUDIES

Even though the analytical method outlined above is quite complicated, it makes use of the existing experimental data on three-dimensional structures of reference proteins to predict possible three-dimensional configurations of peptides of known amino acid sequences. However, much more work will be required to understand how proteins fold into their active tertiary structures with the information content of their amino acid sequences (Anfinsen, 1973). In some simple cases, the amino acid sequences of relatively small proteins in the proper environment may be sufficient to determine their folding. For larger proteins, they may require the interaction of other proteins or organelles. Theoretical analysis for the latter situation will be very complicated.

REFERENCES

Anfinsen CB (1973) Principles that govern the folding of protein chains. *Science,* **181**, 223-230.

Appella E, McIntire KR and Perham RN (1967) Lambda Bence Jones proteins of the mouse: chemical and immunological characterization. *J. Mol. Biol.,* **27**, 391-394.

Chothia C and Lesk AM (1987) Canonical structures for the hypervariable regions of immunoglobulins. *J. Mol. Biol.,* **196**, 901-917.

Chou PY and Fasman GD (1974) Prediction of protein conformation. *Biochemistry,* **13**, 222-245.

Dickerson RE and Geis I (1969) *The Structure and Action of Proteins.* W. A. Benjamin, Inc., Menlo Park, CA.

Honig B, Kabat EA, Katz L, Levinthal C and Wu TT (1973) Model-building of neurophyophyseal hormones. *J. Mol. Biol.,* **80**, 277-295.

Hovis JG (1982) Prediction of a three dimensional structure for the MOPC104E-α(1–3) dextran antibody-antigen complex. *Ph. D. Dissertation,* Northwestern University.

Hovis JG and Wu TT (1985) Prediction of the side chain configurations. A possible three-dimensional structure of the MOPC104E-alpha(1–3) dextran antibody-antigen complex. In *New Methodologies in Stdueis of Protein configuration,* Ed. T. T. Wu, Van Nostrand Reinhold Co., New York, NY, pp. 142-164.

Epp O, Lattman EE, Schiffer M, Huber R and Palm W (1975) The molecular structure of a dimer composed of the variable portions of the Bence-Jones protein REI refined at 2.0 A resolution. *Biochemistry,* **14**, 4943-4952.

Kabat EA, Wu TT, Perry H, Gottesman K and Foeller C (1991) *Sequences of Proteins of Immunological Interest, 5th Edition,* NIH Publication No. 91-3242, Bethesda, MD.

Kar L, Sherman SA and Johnson ME (1994) Comparison of protein structure in solution using local conformations derived from NMR data: application to cytochrome c. *J. Biomol. Struct. Dyn.,* **12**, 527-558.

Kehry M, Sibley C, Fuhrman J, Schilling J and Hood LE (1979) Amino acid sequence of a mouse immunoglobulin mu chain. *Proc. Nat. Acad. Sci. USA,* **76**, 2932-2936.

Poljak RJ, Amzel LM, Avey HP and Becka LN (1972) Structure of Fab' New at 6 A resolution. *Nat. New Biol.,* **235**, 137-140.

Schepers G, Blatt Y, Himmelspach K and Pecht I (1978) Binding site of a dextran-specific homogeneous IgM: thermodynamic and spectroscopic mapping by dansylated oligosaccharides. *Biochemistry,* **17**, 2239-2245.

Schiffer M and Edmundsen AB (1967) Use of helical wheels to represent the structures of preoteins and to identify segments with helical potential. *Biophys. J.*, 7, 121-135.

Stanford JM (1979) Prediction of the three-dimensional conformation of the variable region of immunoglobulin MOPC-315. *Ph. D. Dissertation*, Northwestern University.

Stanford JM and Wu TT (1979) A predictive method for determining possible three-dimensional folding of immunoglobulin backbones around antibody combining sites. *J. Theoret. Biol.*, **88**, 421-439.

Stevens PW and Wu TT (1985) Prediction of protein backbone conformations. A possible three-dimensional structure of ribitol dehydrogenase. In *New Methodlogies in Studies of Protein Configuration*, Ed. T. T. Wu, Van Nostrand Reinhold Co., New York, NY, pp. 116-141.

Webster DM and Rees AR (1995) Molecular modeling of antibody combining sites. *Methods Mol. Biol.*, **51**,17-49.

Wu TT, Fitch WM and Margoliash E (1974) Prediction of secondary structure of protein amino acid sequences. *Ann. Rev. Biochem.*, **43**, 539-566.

EXERCISE

Plot the backbone atomic coordinates of MOPC-315, and rotate around the x-, y- and z-axis separately using the rotation matrix in the Appendix.

CHAPTER 7

STRUCTURES OF POLYNUCLEOTIDES

INTRODUCTION

Similar to polypeptides, periodic structures of polynucleotides can also be analyzed by steric hindrance studies. The basic building block is a nucleotide. For RNA, i.e. ribonucleic acid, there are four ribonucleotides with different bases: adenine (A), cytosine (C), guanine (G) and uracil (U). For DNA, i.e. deoxyribonucleic acid, there are four deoxyribonucleotides with different bases: adenine (A), cytosine (C), guanine (G) and thymine (T). The ribose ring is more or less flat, similar to the peptide bond in polypeptides, although it can assume several different configurations. The side chain configurations of nucleotides are specified by one single bond free to rotate. Unlike the polypeptide backbone where each amino acid residue has only two single bonds with rotational degree of freedom, designated by the ϕ and ψ angles, polynucleotide backbone has five degrees of rotational freedom for each nucleotide as shown in Fig. 7-1 (Metzler, 1977). Thus, to define a periodic structure needs at least a five-dimensional space. For this reason, very few studies on periodic structures of polynucleotides have been carried out. It is a wide-open field in molecular biology for detailed theoretical analysis.

In this Chapter, some historical attempts to study polynucleotide structures are summarized. Nomenclatures used by various authors are different. It is hoped that once the periodic structures of polynucleotides are understood, detailed three-dimensional structures of RNA and DNA can be analyzed. Tremendous amount experimental data are already available for relatively short RNA and DNA segments. However, even the packing of DNA inside a λ–phage is beyond comprehension.

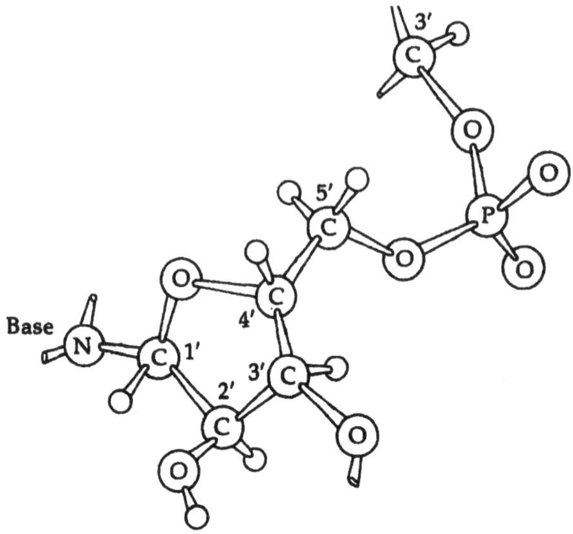

Figure 7-1. Configuration of a nucleotide unit.

PERIODIC STRUCTURES

Similar to polypeptides, Pauling and Corey (1953) also proposed that DNA could exist in periodic structures. Many other periodic structures were subsequently suggested, and several of these will be discussed in the next Chapter. Some are right-handed, and others left-handed. However, unlike polypeptides, the basic building block has many more single bonds free to rotate. As a result, periodic structures of polynucleotides can not be visualized easily in a two-dimensional Ramachandran plot. There is also no simple paper model as that for the peptide backbone illustrated in Chapter 5. Instead, a minimum of a five-dimensional space will be required for the backbone, together with possible distortions of the ribose ring. Over the years, virtual angles have been suggested. However, it is usually very difficult to reduce any description in a five-dimensional space to a two-dimensional space.

On the other hand, the side chains of polynucleotides have much simpler configurations as compared to those of polypeptides. The four bases of both RNA and DNA are all flat structures, and are linked to the ribose ring by a single glycosidic bond. This bond will be analyzed first.

GLYCOSIDIC BOND

Similar to the minimum allowed contact distances between non-bounded atoms, another approach of studying steric hindrance is to use the atomic and van der Waal radii of various atoms as listed in Table 7-1.

Table 7-1. Atomic and van der Waal radii of H, C, N, O and P.

Atom	Atomic radius (A)	van der Waal radius (A)
H	0.30	1.2
C	0.77	1.6
N	0.75	1.5
O	0.74	1.4
P	1.10	1.9

The van der Waal radii listed above are on the larger side. Some authors assume that for H atom can be as small as 0.9 A. For numerical evaluations, we calculate a quantity, designated as tolerance Δ, among any two sets of non-bounded atoms i and j (Wu, 1968):

$$\Delta = \min_{i,j} \{\text{distance between atoms i and j} - \text{van der Waal radius of atom i} - \text{van der Waal radius of atom j}\}.$$

Atom i denotes atoms on one side of the single bond in consideration, and atom j those on the other side of the single bond. For $\Delta > 0$, there is no steric hindrance. For $\Delta < 0$, it does not necessarily indicate disallowed configurations, since the minimum allowed van der Waal radii may be smaller than those listed in Table 7-1, by as much as 0.3 A.

The rotation around the glycosidic bond is denoted by χ in Fig. 7-1, but was by θ_1 as illustrated in Fig. 7-2 (Wu, 1968).

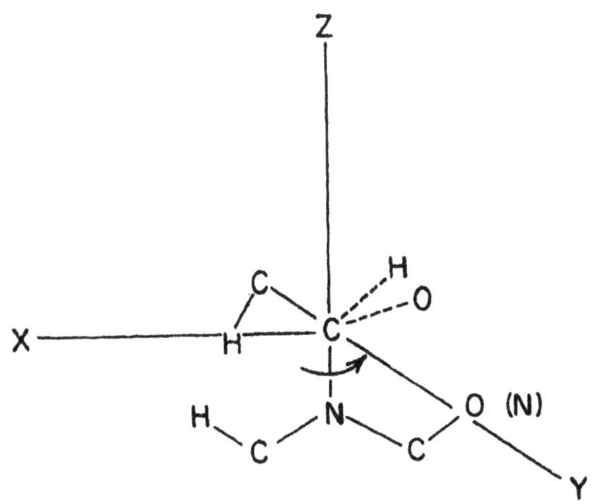

Figure 7-2. Definition of θ_1, rotation around the glycosidic bond.

The ring structure of the pyrimidine or purine base lies in the x,z-plane and below the x,y-plane. It is held stationary. The glycosidic bond lies along the negative z-axis, with C_1' of the ribose ring at the origin. At $\theta_1 = 0^0$, all of the solid bonds shown in Fig. 7-2 are in the x,z-plane. The angle θ_1 is defined as the angle between the x,z-plane and the plane formed by the glycosidic bond and the C_2'–C_1' bond of the ribose ring. The right-hand rule is used to define the positive direction of θ_1 for rotation around the z-axis. The torsion angle defined by Donohue and Trueblood (1960) is roughly equal to $240^0 - \theta_1$. Only those atoms shown in Fig. 7-2 are considered for steric hindrance calculations.

Tolerance Δ is calculated for pyrimidine and purine bases separately, as illustrated in Figs. 7-3 and 7-4.

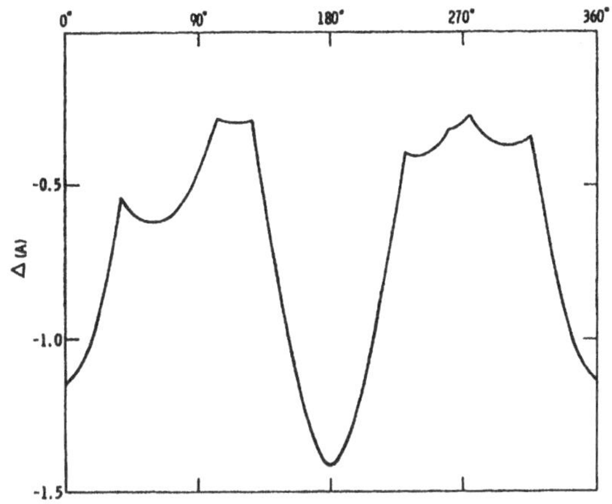

Figure 7-3. Tolerance Δ as a function of θ_1 for pyrimidine base.

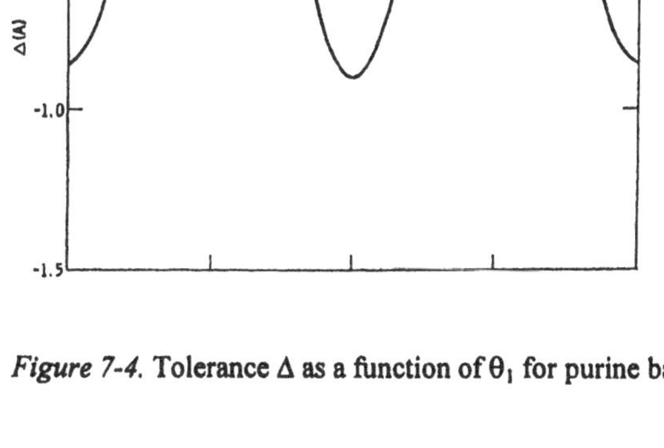

Figure 7-4. Tolerance Δ as a function of θ_1 for purine base.

The negative values of tolerance Δ in Figs. 7-3 and 7-4 suggest that the van der Waal radii of various atoms listed in Table 7-1 should be reduced somewhat, especially that for H atom. However, roughly, we have possible values of θ_1 for pyrimidine and purine in the ranges:

pyrimidine: 95^0 to 130^0, and 230^0 to 315^0;

purine: 40^0 to 160^0, and 220^0 to 320^0.

The value of 100^0 corresponds to the *syn* configuration suitable for the Hoogsteen pairing (Hoogsteen, 1963), and that of 280^0 to the *anti* configuration suitable for the Watson-Crick pairing (Watson and Crick, 1953). Indeed, especially for purine bases, there is a fairly large amount of allowable rotation around the glycosidic bond.

FIVE DEGREES OF FREEDOM OF THE BACKBONE

Rotations around the five single bonds (Metzler, 1977) between two successive phosphorous atoms as shown in Fig. 7-1 have designations overlapping with those for the polypeptide backbone. To avoid such confusion, earlier in 1968, they were designated as θ_2, θ_3, θ_4, θ_5 and θ_6 (Wu, 1968). The C_4'–C_3' bond will be considered separately due to the restrictions imposed by the ribose ring structure.

The angle θ_2 is defined in Fig. 7-5, for the rotation around the C_4'–C_5' bond. The C_4' atom is positioned at the origin, and the C_4'–C_5' bond along the negative z-axis. The CH_2O group of the sugar lies below the x,y-plane, and is held stationary with the O atom in the x,z-plane. The ribose ring is then rotated around the z-axis in the positive direction, with the right-hand rule. At $\theta_2 = 0^0$, all solid bonds in Fig. 7-5 are in the x,z-plane. The angle θ_2 is the angle between the x,z-plane and the plane formed by the C_3'–C_4' and C_4'–C_5' bonds. Again for calculation of tolerance Δ, only those atoms shown in Fig. 7-5 are included. Tolerance as a function of θ_2 is plotted in Fig. 7-6. The least hindered values of θ_2 are around 120^0 and 240^0. Most of the values of θ_2 seem to be allowed, except possible between 130^0 to 230^0.

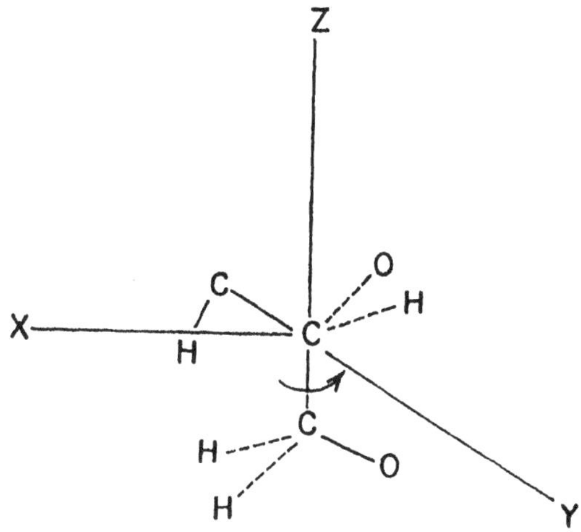

Figure 7-5. Definition of θ_2, rotation around C_4'–C_5' bond.

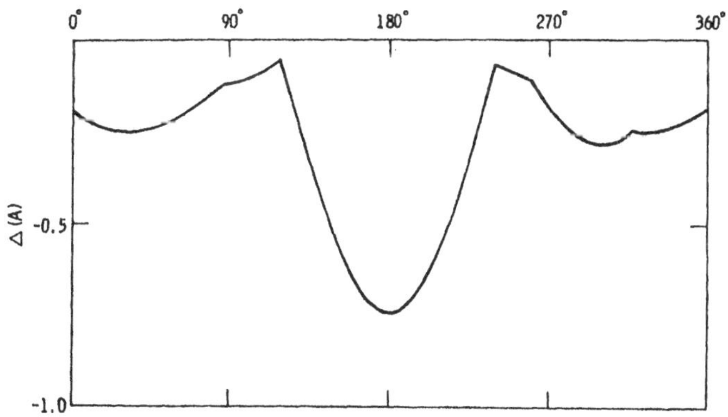

Figure 7-6. Tolerance Δ as a function of θ_2.

The angles θ_3 and θ_4 define the rotations around the O–C_5' bond and the bond between that O atom and the P atom (Figs. 7-7 and 7-8).

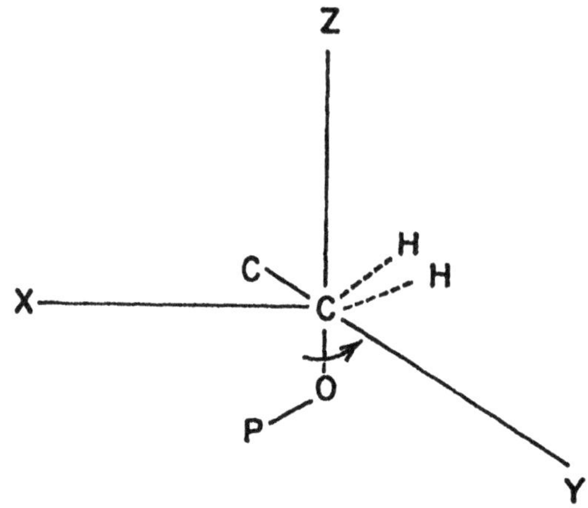

Figure 7-7. Definition of θ_3, rotation around the O–C_5' bond.

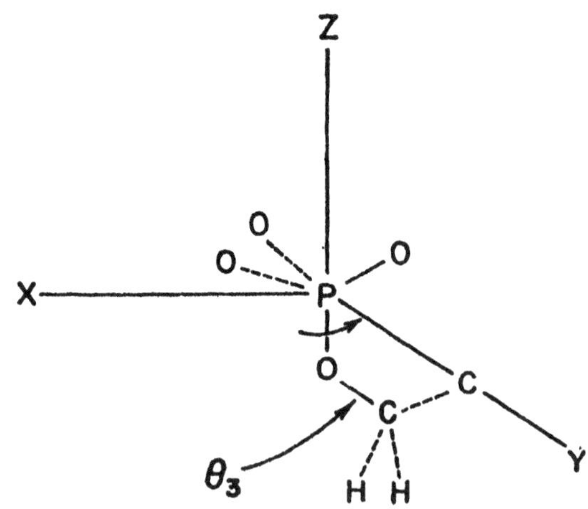

Figure 7-8. Definition of θ_4, rotation around the O–P bond on the 5'-end.

In Fig. 7-7, the C_5' atom is placed at the origin, and the C_5'–O bond along the negative z-axis. The P atom is in the x,z-plane. The angle θ_3 is the angle between the x,z-plane and the plane formed by the C_4'–C_5' bond and the C_5'–O bond. In Fig. 7-8, the P atom is placed at the origin, and the P–O bond on the 5'-end along the negative z-axis with the C_5' atom in the x,z-plane. The angle θ_4 is the angle between the x,z-plane and the plane defined by the (C_3'–)O–P and the P–O(–C_5') bonds. In both cases, the positive θ_3 and θ_4 angle follow the right-hand rule along the z-axis. Due to the relatively large van der Waal radius of the P atom, the tolerance Δ is calculated for various values of θ_3 for the P atom only and illustrated in Fig. 7-9. As expected, when the P atom is opposite to the two H atoms on C_5', i.e. $\theta_3 = 180^0$, there is least steric hindrance. Other locations for low tolerance are around θ_3 of 110^0 and 290^0.

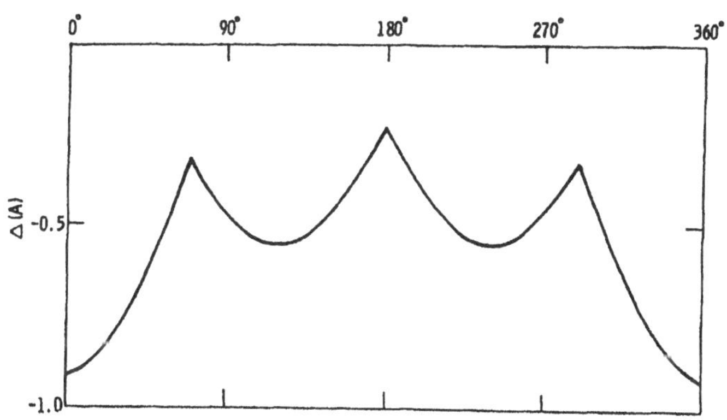

Figure 7-9. Tolerance as a function of θ_3 for P atom only.

Since the variation of θ_3 from 180^0 to 360^0 is similar to that from 180^0 to 0^0, only the latter range is considered together with the θ_4 for tolerance Δ calculation involving other atoms. The result is shown in Fig. 7-10, with the P atom excluded. The region with tolerance greater than zero ($\Delta > 0$) is stippled.

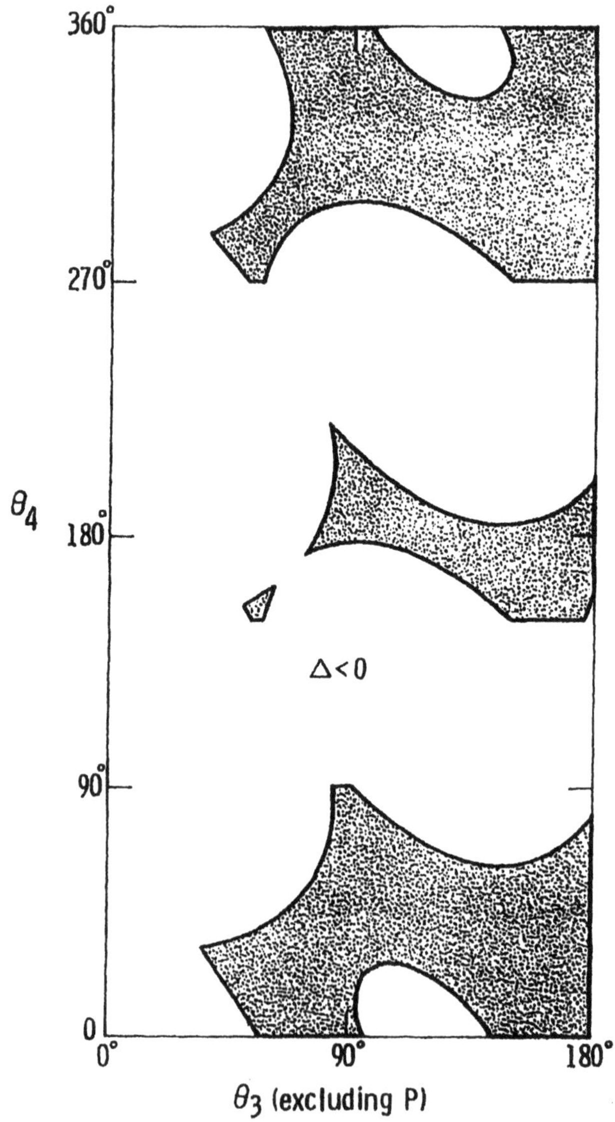

Figure 7-10. Allowable values of angles θ_3 and θ_4 with the omission of the P atom.

Similarly, angles θ_5 and θ_6 are defined as the rotation around P–O and O–C_3' bonds as illustrated in Figs. 7-11 and 7-12.

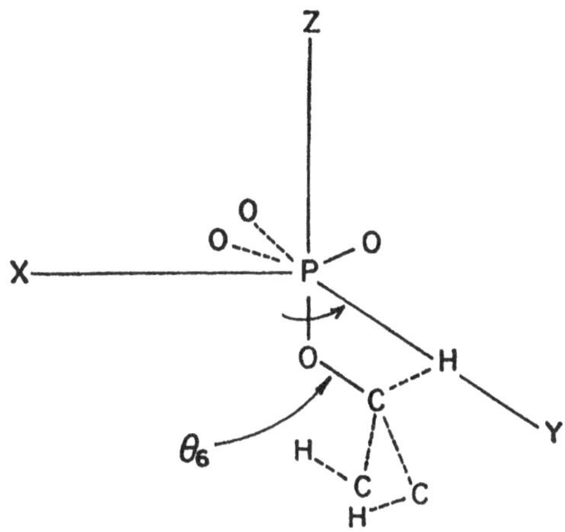

Figure 7-11. Definition of θ_5, rotation around the P–O bond on the 3'-end.

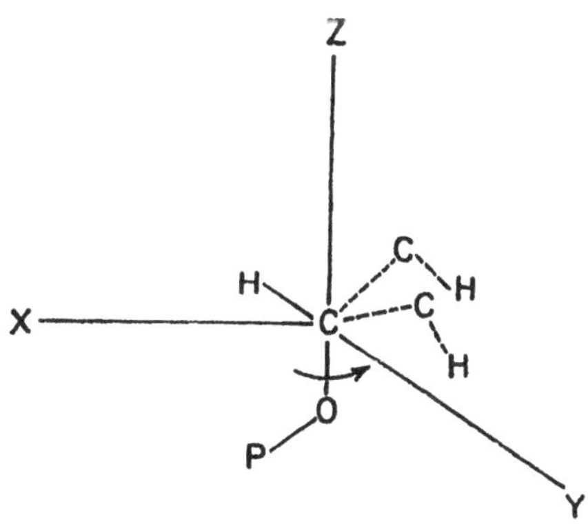

Figure 7-12. Definition of θ_6, rotation around the O–C_3' bond.

In Fig. 7-11, the P atom is placed at the origin, and the P–O(–C_3') along the negative z-axis. The other P–O(–C_5') bond is in the x,z-plane at $\theta_5 = 0°$. Angle θ_3 is the angle between the x,z-plane and the plane defined by these two P–O bonds. In Fig. 7-12, the C_3' atom is placed at the origin, and the C_3'–O and the O–P bonds are in the x,z-plane. Angle θ_6 is the angle between the x,z-plane and the plane defined by the H_3'–C_3' and the C_3'–O bonds. In both cases, the positive value follows the right-hand rule along the z-axis. Again, due to the relative large van der Waal radius of the P atom, tolerance Δ is plotted again θ_6 for P atom only in Fig. 7-13.

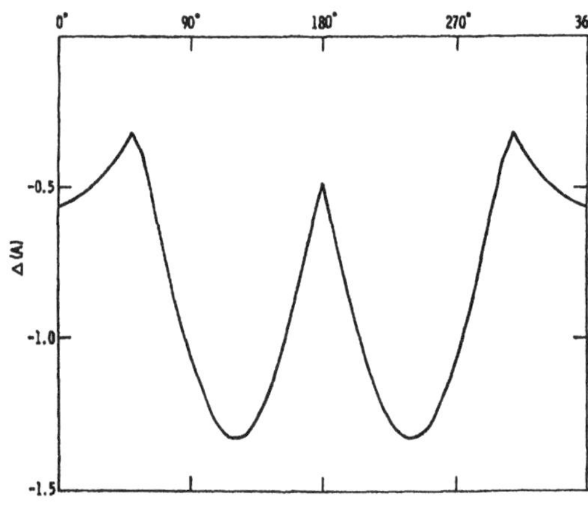

Figure 7-13. Tolerance Δ as a function of θ_6 for P atom only.

Similar to the situation for the combination of θ_3 and θ_4, tolerance Δ can also be plotted for the combination of θ_5 and θ_6 excluding the P atom. Symmetry of θ_6 still exists for DNA. However, the presence of the OH group attached to the C_2' atom on RNA makes this symmetry no longer valid (Murthy *et al.*, 1999). Fig. 7-14 illustrates tolerance as a function of θ_5 and θ_6 for DNA. Thus, it is clear that the 3'-end of the nucleotide unit is more restricted due to steric hindrance than the 5'-end.

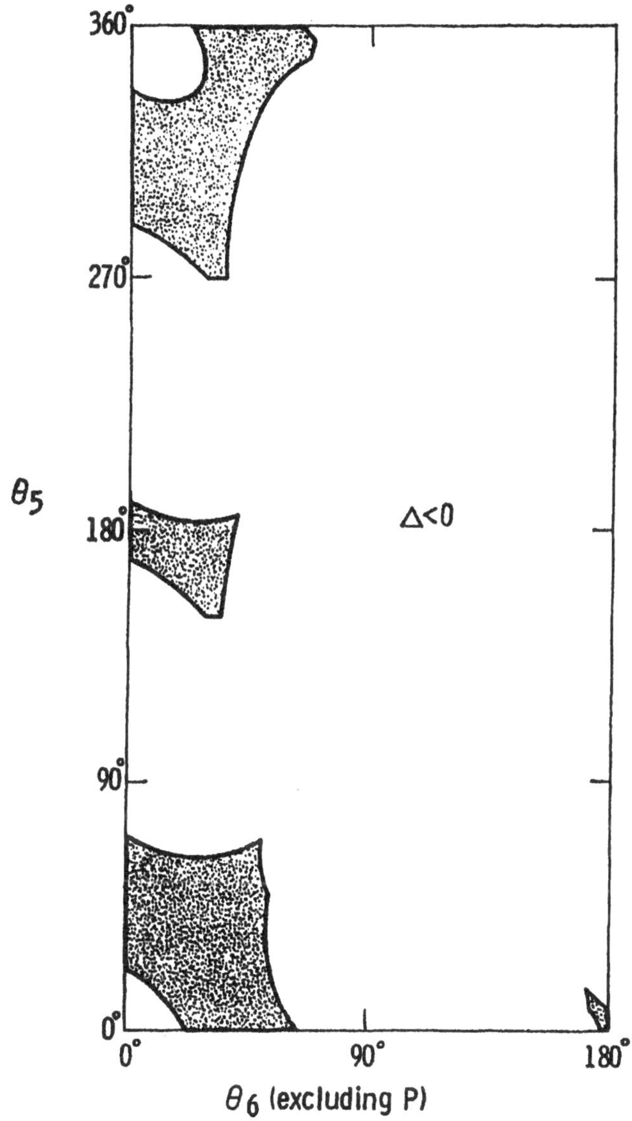

Figure 7-14. Allowable values of angles θ_5 and θ_6 with the omission of the P atom.

RIBOSE RING STRUCTURE

As illustrated in Fig. 7-1, the ribose ring structure is not planar. The C_2' atom or the C_3' atom can be displaced out of the plane towards the C_5' atom. Thus, even though the C_3'–C_4' bond is not free to rotate, it can assume several different positions, commonly referred to as pucker (Sarma, 1980). Thus, it provides additional complication to the backbone structure of polynucleotides, similar to the idea that peptide bonds in polypeptides may not be completely planar.

Therefore, to understand the complete nature of periodic or secondary structures of DNA and RNA requires detailed calculations of all sterically allowed configurations. Such calculations must be carried out in the five-dimensional space defined by angles θ_2, θ_3, θ_4, θ_5 and θ_6 , together with the glycosidic bond defined by θ_1 and the puckering phase and amplitude of the ribose ring structure (Marzec and Day, 1993; Murthy *et al.*, 1999). As mentioned before, various authors use different designations for these angles. Some of them are listed in Table 7-2.

Table 7-2. Notations used to designate rotations around single bonds of a nucleotide.

Bond	Wu, 1968	Metzler, 1977	Murthy *et al.*, 1999
P–O	θ_4	ψ	α
O–C_5'	θ_3	θ	β
C_5'–C_4'	θ_2	ξ	γ
C_4'–C_3'		σ	P,A
C_3'–O	θ_6	ω	ε
O–P	θ_5	ϕ	ζ
glycosidic	θ_1	χ	χ

Unified notations, as well as conventions of angles, can then provide a careful study of polynucleotides conformations (Murthy *et al.*, 1999) similar to the Ramachandran plot for polypeptides.

VIRTUAL ANGLES

In the hope of reducing the complexity of the polynucleotide backbone configuration to a two-dimensional plot, various virtual angles have been proposed during the past 30 years. Olson and Flory (1972) proposed to consider two virtual bonds, one from P to C_5' and the other from C_5' to the other P of the repeating nucleotide unit. They were therefore able to use a two-dimensional representation of two virtual angles for the rotations around these two virtual bonds. Duarte and Pyle (1998) suggested a slight variation, using two virtual bonds from P to C_4' and from C_4' to the other P. Whether a two-dimensional description can substitute the real situation of a five- or six-dimensional space is open for discussion.

FUTURE ANALYSIS

While the polynucleotide configurations are more complicated than those of polypeptides, it is essential to analyze them in complete detail regardless of whether multi-dimensional space will be required. Once their periodic or secondary structures are thoroughly understood, detailed comparison with experimental data can provide valuable information about the three-dimensional folding of DNA and RNA. In general, due to the presence of five single bonds with some degree of freedom to rotate for each nucleotide, the backbone structure of DNA and RNA may be quite flexible. With the rapid development of experimental study of tertiary structures of RNA molecules and short DNA segments, it may soon be possible to understand how DNA is packaged tightly inside the λ–phage heads, into chromosomes which can duplicate without tangling, etc.

REFERENCES

Donohue J and Trueblood KN (1960) Base pairing in DNA. *J. Mol. Biol.*, **2**, 363-371.

Duarte CM and Pyle AM (1998) Stepping through an RNA structure: A novel approach to conformational analysis. *J. Mol. Biol.*, **284**, 1465-1478.

Hoogsteen K (1963) The crystal and molecular structure of a hydrogen-bonded complex between 1-methylthymine and 9-methyladenine. *Acta Cryst.*, **16**, 907-916.

Marzec CJ and Day LA (1993) An exact description of five-membered ring configurations. I. Parameterization via an amplitude S, an angle Γ, the pseudorotation amplitude q and the phase angle P, and the bond lengths. *J. Biomol. Struct. Dyn.*, **10**, 1091-1123.

Metzler DE (1977) *Biochemistry*. Academic Press, New York, NY.

Murthy VL, Srinivasan R, Draper DE and Rose GD (1999) A complete conformational map for RNA. *J. Mol. Biol.*, **291**, 313-327.

Olson WK and Flory PJ (1972) Spatial configurations of polynucleotide chains. I. Steric interactions in polynucleotides: a virtual bond model. *Biopolymers,* **11**, 1-23.

Pauling L and Corey RB (1953) A proposed structure for the nucleic acids. *Proc. Nat. Acad. Sci. USA,* **39**, 84-97.

Sarma RH (1980) *Nucleic Acid Geometry and Dynamics.* Pergamon Presee, New York, NY.

Watson JD and Crick FHC (1953) The structure of DNA. *Cold Spring Harbor Symp. Quant. Biol.*, **18**, 123-131.

Wu TT (1968) Periodic conformations of deoxyribonucleic acids. *Bull. Math. Biophys.*, **30**, 687-700.

EXERCISE

Calculate the sterically allowed configurations as a function of the rotation around the glycosidic bond for the four bases separately, using the minimum contact distances between non-bounded atoms listed in Chapter 5. Try to find the bond distances and bond angles for the ribose ring and A, C, G, T from standard textbooks.

CHAPTER 8

FIBER X-RAY DIFFRACTION AND DNA DOUBLE HELIX

INTRODUCTION

Fiber X-ray diffraction has been used to determine the secondary structure of DNA. In 1953, Franklin and Gosling obtained the first useful experimental result. To analyze their diffraction picture in detail, we need to examine its near forward region shown in Fig. 8-1, up to the fourth layer line. That picture is symmetric left-to-right and top-to-bottom:

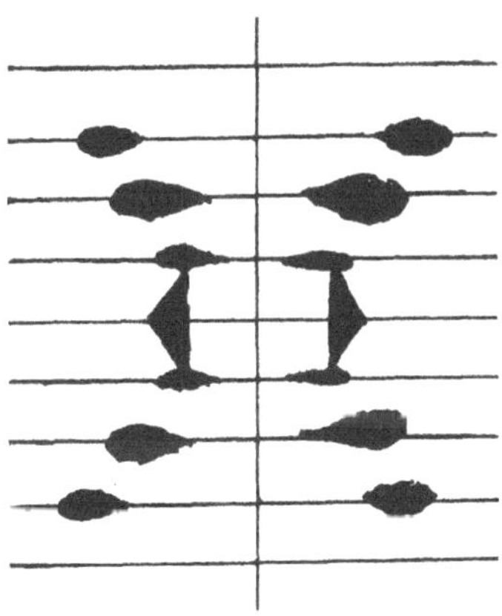

Figure 8-1. Sketch of the fiber X-ray diffraction picture of the sodium salt of DNA obtained Franklin and Gosling (1953), including only up to the fourth layer line.

In the same issue of Nature, two other important papers were also published: one by Watson and Crick (1953) showing the double helix model, and the other by Wilkins, Stokes and Wilson (1953) including the following paragraph:

"It must first be decided whether the structure consists of essentially one helix giving an intensity distribution along the layer lines corresponding to J_1, J_2, J_3 ..., or two similar co-axial helices of twice the above size and relatively displaced along the axis a distance equal to half the pitch giving J_2, J_4, J_6 ..., or three helices, etc. Examination of the width of the layer-line streaks suggests the intensities correspond more closely to J_1^2, J_2^2, J_3^2 than to J_2^2, J_4^2, J_6^2 ..."

The J_n is the nth order of Bessel function (see, for example, Whittaker and Watson, 1915). It is therefore important for us to understand the properties of Bessel function, and also the reason why helices give X-ray diffraction patterns related to Bessel function (see, for example, Cochran, Crick and Vand, 1952).

SOME SIMPLE PROPERTIES OF J_n

Bessel function J_n can be defined by the following generating function (see, for example, Whittaker and Watson 1915):

$$G(x,t) = \exp\,[x(t - 1/t)/2] = \Sigma\, t^n J_n(x).$$

By taking partial derivatives of G(x,t) with respect to x and t separately, we get:

$$(t - 1/t)/2\, \Sigma\, t^n J_n(x) = \Sigma\, t^n dJ_n(x)/dx,$$

and $$x(1 + 1/t^2)/2 \sum t^n J_n(x) = \sum n\, t^{n-1} J_n(x).$$

Collecting terms with the same t^n, we derive the following two recurrence relations:

$$[J_{n-1}(x) - J_{n+1}(x)]/2 = dJ_n(x)/dx.$$

and $$x[J_n(x) + J_{n+2}(x)]/2 = (n+1)\, J_{n+1}(x).$$

The second equation can also be written as:

$$x[J_{n-1}(x) + J_{n+1}(x)]/2 = n\, J_n(x).$$

On eliminating $J_{n-1}(x)$ and $J_{n+1}(x)$ from these two equations, we can verify that $J_n(x)$ is one of the solutions of the following Bessel differential equation:

$$d^2y/dx^2 + 1/x\, dy/dx + (1 - n^2/x^2)\, y = 0,$$

The other solution is denoted by $Y_n(x)$ which is singular at $x = 0$.

If we consider t as the complex variable, Cauchy's integral theorem can be applied to obtain the value of contour integral around a unit circle, i.e. $t = \exp[i\theta]$, in the counter-clock-wise direction of $G(x,t)/t^{n+1}$. Since

$$(t - 1/t)/2 = i \sin\theta,$$

we get,

$$2\pi J(x) = \int_{\theta_o}^{\theta_o+2\pi} \exp\,[ix \sin\theta - in\theta]\, d\theta.$$

where θ_o is any angle. Values of $J_n(x)$ can be found in tables of mathematical functions (see, for example, Abramowitz and Stegun, 1965). Those of $J_1(x)$, $J_2(x)$, $J_3(x)$ and $J_4(x)$ in the range of x between 0.0 to 8.0 are listed in Table 8-1. Only five decimal places are included. Bessel functions are also available on various computers.

Table 8-1. Values of $J_n(x)$ for n = 1, 2, 3 and 4, and 0.0 < x < 8.0.

x	$J_1(x)$	$J_2(x)$	$J_3(x)$	$J_4(x)$
0.0	0.00000	0.00000	0.00000	0.00000
0.2	0.09950	0.00498	0.00017	0.00000
0.4	0.19603	0.01973	0.00132	0.00007
0.6	0.28670	0.04367	0.00440	0.00033
0.8	0.36884	0.07582	0.01025	0.00103
1.0	0.44005	0.11490	0.01956	0.00248
1.2	0.49829	0.15935	0.03287	0.00502
1.4	0.54195	0.20736	0.05050	0.00906
1.6	0.56990	0.25697	0.07252	0.01500
1.8	0.58152	0.30614	0.09880	0.02320
2.0	0.57672	0.35283	0.12894	0.03400
2.2	0.55596	0.38506	0.16233	0.04765
2.4	0.52019	0.43098	0.19811	0.06431
2.6	0.47082	0.45897	0.23529	0.08401
2.8	0.40971	0.47769	0.27270	0.10667
3.0	0.33906	0.48609	0.30906	0.13203

3.2	0.26134	0.48353	0.34307	0.15972
3.4	0.17923	0.46972	0.37339	0.18920
3.6	0.09547	0.44481	0.39876	0.21980
3.8	0.01282	0.40930	0.41803	0.25074
4.0	-0.06604	0.36413	0.43017	0.28113
4.2	-0.13865	0.31053	0.43439	0.31003
4.4	-0.20278	0.25009	0.43013	0.33645
4.6	-0.25655	0.18459	0.41707	0.35941
4.8	-0.29850	0.11605	0.39521	0.37796
5.0	-0.32758	0.04657	0.36483	0.39123
5.2	-0.34322	-0.02172	0.32652	0.39847
5.4	-0.34534	-0.08670	0.28113	0.39906
5.6	-0.33433	-0.14638	0.22978	0.39257
5.8	-0.31103	-0.19895	0.17382	0.37877
6.0	-0.27668	-0.24287	0.11477	0.35764
6.2	-0.23292	-0.27688	0.05428	0.32941
6.4	-0.18164	-0.30007	-0.00591	0.29453
6.6	-0.12498	-0.31192	-0.06406	0.25368
6.8	-0.06522	-0.31228	-0.11847	0.20774
7.0	-0.00468	-0.30142	-0.16756	0.15780
7.2	0.05433	-0.27998	-0.20987	0.10509
7.4	0.10963	-0.24897	-0.24420	0.05097
7.6	0.15921	-0.20970	-0.26958	-0.00313
7.8	0.20136	-0.16378	-0.28535	-0.05572
8.0	0.23464	-0.11299	-0.29113	-0.10536

In order to reconcile with the discussion of Wilkins, Stokes and Wilson (1953), we have plotted $J_1^2(x)$, $J_2^2(x)$, $J_3^2(x)$ and $J_4^2(x)$ vs. x in Fig. 8-2 with the values in Table 8-1.

Their first peaks gradually reduce in height, and their half widths increase only slightly. In order to differentiate them, the most significant feature is their locations, or the values of x at their first maximums. They are roughly around 1.8, 3.0, 4.2 and 5.4 respectively. As clearly indicated in Fig. 8-2, the widths of their first peaks cannot tell them apart at all.

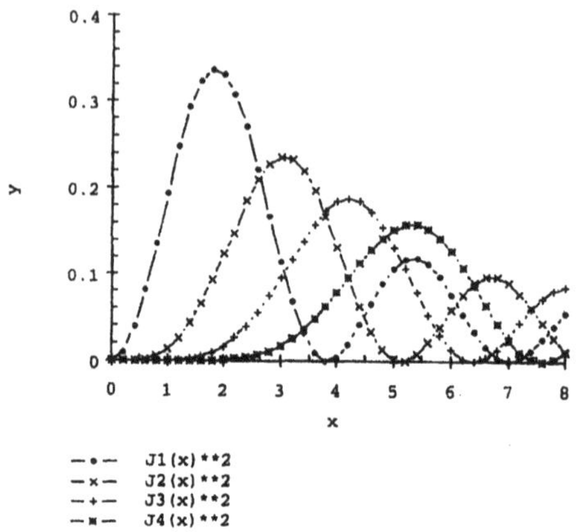

Figure 8-2. Plots of $J_1^2(x)$, $J_2^2(x)$, $J_3^2(x)$ and $J_4^2(x)$ for $0 < x < 8$.

X-RAY DIFFRACTION PATTERN OF HELICAL LINE

In three-dimension, the equations which define a right-handed helix are:

$$x = r\cos(2\pi z/P),$$

and

$$y = r\sin(2\pi z/P),$$

where r is the radius of the helix, and P its pitch. The helical axis is along the z-axis (Fig. 8-3). For a left-handed helix, the only change is to insert a minus sign in front of r for the equation of y, i.e.

$$x = r\cos(2\pi\zeta/P),$$

and
$$y = -r\sin(2\pi\zeta/P).$$

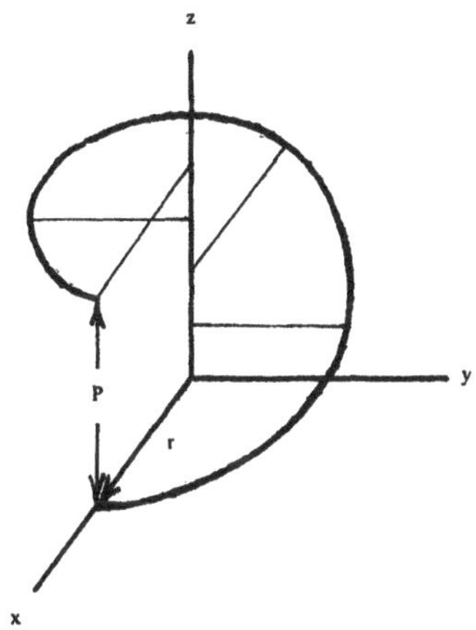

Figure 8-.3. A right-handed helix of radius r and pitch P.

Consider an X-ray beam from a source hitting a specimen. The electrons in the specimen are set into vibration around their equilibrium positions and emit diffracted beams of the same wave length. The diffracted beams will interfere with each other, according to the Bragg's law of diffraction, and are recorded on a film. For two electrons, located at A and B, in the specimen (Fig. 8-4), let $\mathbf{s}_o$ denote the unit vector in the incident direction, and $\mathbf{s}$ the unit vector in the diffracted direction. The diffracted wave for electron at B will lag behind that for electron at A by a distance equal to AC – DB. If the position vector from A to B is denoted by $\mathbf{x}$, then

$$AC = \mathbf{x} \cdot \mathbf{s},$$

$$DB = \mathbf{x} \cdot \mathbf{s}_o,$$

and
$$AC - DB = \mathbf{x} \cdot (\mathbf{s} - \mathbf{s}_o).$$

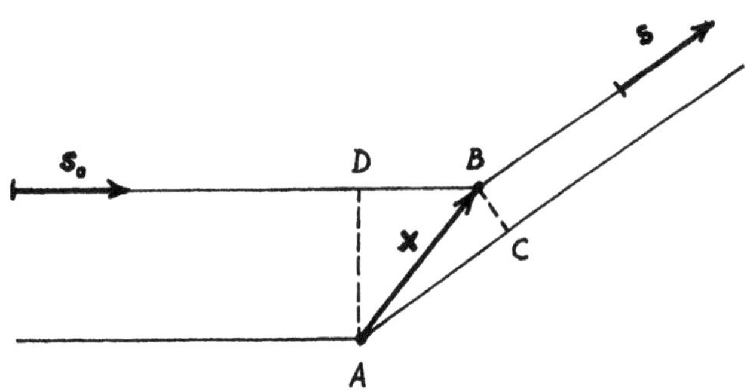

Figure 8-4. Basic principle of X-ray diffraction for two electrons located at A and B.

To add the amplitudes of the diffracted beams from these two electrons in the phase plane vectorially, we need to calculate the phase difference from the path difference by multiplying with $2\pi/\lambda$, where λ is the wave length of the X-ray. Thus, if the amplitude of the diffracted beam from electron A is denoted by $\mathbf{A}_o$, then that from electron B is $\mathbf{A}_o \exp[2\pi i\, \mathbf{x} \cdot (\mathbf{s} - \mathbf{s}_o)/\lambda]$, in the direction of $\mathbf{s}$. The X-ray film then can be used to determine the reciprocal space coordinates defined as:

$$\mathbf{h} = (\mathbf{s} - \mathbf{s}_o)/\lambda.$$

Just like the physical space coordinates defined by $\mathbf{x} = (x,y,z)$, we also have $\mathbf{h} = (\xi,\eta,\zeta)$, and

$$\mathbf{x} \cdot \mathbf{h} = x\xi + y\eta + z\zeta.$$

If we represent the specimen as electron density distribution, $\rho(\mathbf{x})$, over its volume, the total amplitude of the diffracted beams from all its electrons can be calculated by the following integration:

$$\mathbf{A}(\mathbf{h}) = \mathbf{A}_o \int \rho(\mathbf{x}) \exp[2\pi i\,(x\xi + y\eta + z\zeta)]\, d^3\mathbf{x},$$

where the triple integral indicates integration over the entire volume of the specimen.

For uniformed electron density distribution along a right-handed helix, since x and y are expressed in terms of z, the triple integral can be reduced to a single integral. Also replace ξ and η by:

$$\xi = R\cos\psi,$$

and

$$\eta = R\sin\psi.$$

The final expression of the integral becomes:

$$\int \exp\{2\pi r R i\,[\cos(2\pi z/P)\cos\psi + \sin(2\pi z/P)\sin\psi] + 2\pi i z\zeta\}\,dz.$$

However, since the helix has a pitch P, the Bragg's law of diffraction gives constructive interference only on layer lines determined by

$$P\sin\alpha = n\lambda,$$

where α is the angle between unit vectors **s** and $\mathbf{s}_o$, and n is an integer. In the near forward direction, the layer lines are shown in Fig. 8-1, with specific values of ζ equal to:

$$2\sin(\alpha/2)/\lambda = \sin\alpha/\lambda = n/P,$$

for α being very small. The integration over the entire length of z can be replace by from 0 to P. Together with a change of variable,

$$2\pi z/P - \psi = \pi/2 - \theta,$$

the total amplitude of the diffracted beam, A(**h**), is proportional to:

$$\int_{\psi - 3\pi/2}^{\psi + \pi/2} \exp\{2\pi r R i\sin\theta + in(-\theta + \pi/2 + \psi)\}\,d\theta.$$

Comparing this with the integral representation of the Bessel function, we get that **A(h)** is proportional to (Franklin and Gosling, 1953):

$$J_n(2\pi rR)\exp[in(\pi/2 + \psi)],$$

The intensity on the various layer lines, **I(h)**, being the square of the absolute value of **A(h)**, is thus proportional to $J_n^2(2\pi rR)$. It becomes essential for us to calculate the values of $2\pi rR$ for the various intensity spots on the X-ray diffraction picture, since the peaks of the Bessel functions are located at specific values of its arguments. Only then, we can understand the statement made by Wilkins, Stokes and Wilson (1953).

As pointed out by Franklin and Gosling (1953), the intensity spot on the equatorial layer line (Fig. 8-1) was due to the side-to-side packing of helices. Again, using the Bragg's law of diffraction, the value of R for that spot is roughly equal to 1/2r. Thus, $2\pi rR$ has an approximate value of π or 3.14. Among all $J_n^2(2\pi rR)$, the only one has a peak roughly at that value of its argument around 3.14 is $J_2^2(2\pi rR)$. Width of the $J_n^2(2\pi rR)$, as suggested by Wilkins, Stokes and Wilson (1953), is not a good measure to differentiate them. See Table 8-1 and Fig. 8-2 for details.

While the original X-ray diffraction picture obtained by Franklin and Gosling in 1953 was for sodium salt of DNA fiber at 92% relative humidity, interestingly enough, Langridge and coworkers obtained another picture in 1960 for the lithium salt of DNA fiber at 66% relative humidity as shown in Fig. 8-5.

We shall try to use the mathematics developed above to understand how the DNA double helix can give rise to one of the experimentally obtained diffraction pictures shown in Fig. 8-1 or 8-5. Then we need to give a plausible explanation to the other experimental diffraction picture.

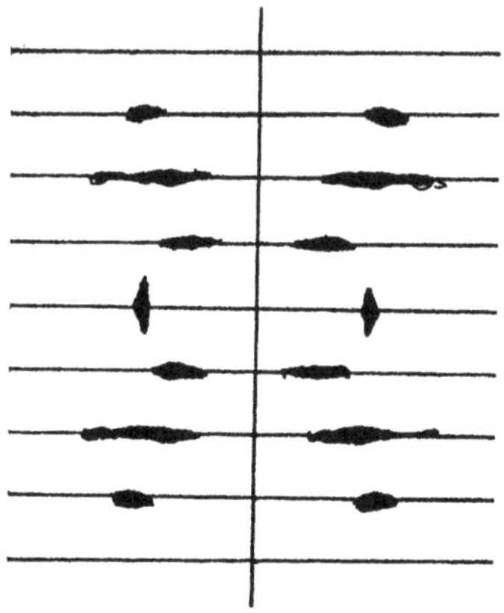

Figure 8-5. Sketch of the fiber X-ray diffraction picture of the lithium salt of DNA obtained by Wilkins' group (1960), including only up to the fourth layer line.

DIFFRACTION PATTERNS OF COMBINATION OF HELICES

Bessel function has well defined properties. The first peak of $J_1^2(2\pi rR)$ is higher than that of $J_2^2(2\pi rR)$, etc., and that the former occurs at a smaller value of $2\pi rR$. In fact, there is a theorem stating that the zeros of successive Bessel function occur between the zeros of the former one. Thus, since the intensity spots on the first layer line are weaker than those on the second layer line for 92% as well as 66% relative humidity (Figs. 8-1 and 8-3), and since the intensity spots on the fourth layer line are missing, the presence of another helix is indicated. In order to determine the relative positions of the two helices, we would consider adding their amplitudes on the first four layer lines in the phase plane (Figs. 8-6 and 8-7).

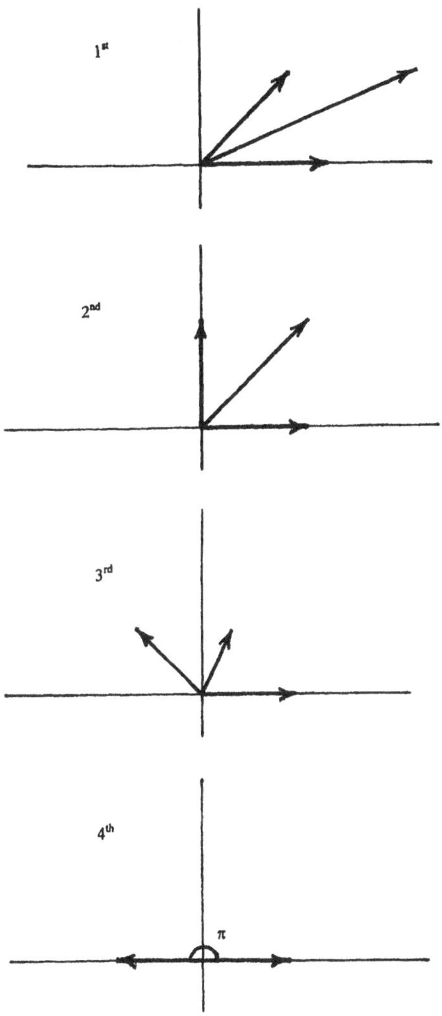

Figure 8-6. The phase difference on the fourth layer line, 4ϕ, is equal to π.

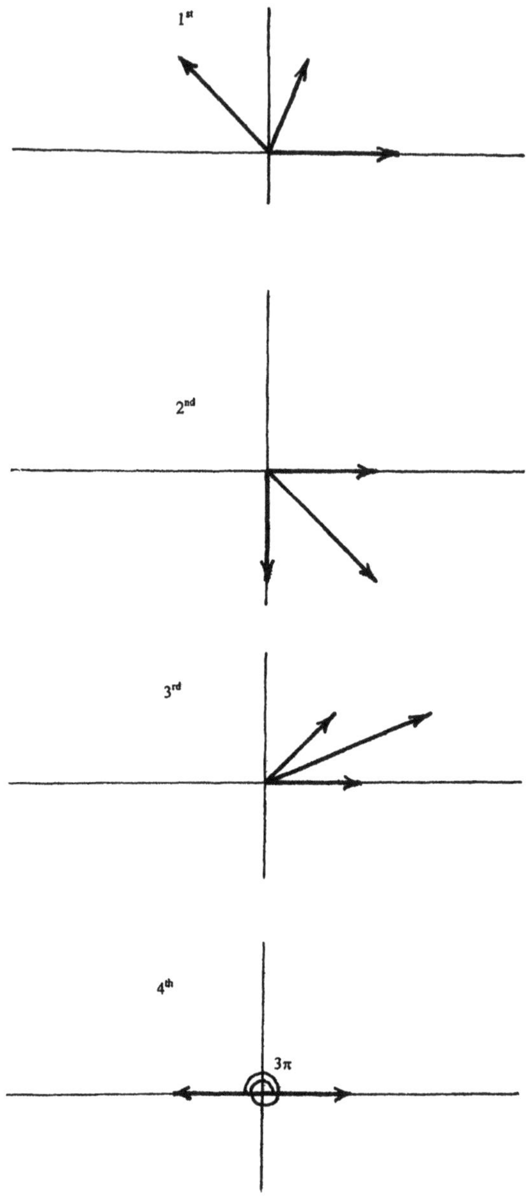

Figure 8-7. The phase difference on the fourth layer line, 4ϕ, is equal to 3π.

The arrows representing the amplitudes on the first four layer lines are scaled to the maximum values of $J_1(x)$, $J_2(x)$, $J_3(x)$ and $J_4(x)$ as listed in Table 8-1.

The absence of any intensity spot on the fourth layer line (Figs. 8-1 and 8-5) suggests that the vectors representing the amplitudes of the two helices must be opposite to each other in the phase plane. Thus, the phase difference on the fourth layer line, 4ϕ, is equal to π, 3π, 5π or 7π. However, the situation for 5π is similar to that of 3π (Fig. 8-5), and 7π to π (Fig. 8-4). For $4\phi = \pi$, or $\phi = \pi/4$, the addition of the two amplitudes on the first layer line (Fig. 8-6) results in a large vector. Thus, the intensities of the spots on the first four layer lines would be very strong, strong, weak and absent, quite different from those observed for DNA fibers (Figs. 8-1 and 8-5). On the other hand, for $4\phi = 3\pi$, or $\phi = 3\pi/4$, the resultant amplitude on the first layer line is much smaller than individual amplitude (Fig. 8-7). Intensities would then be weak, strong, strong and absent on the first four layer lines, as observed experimentally (Figs. 8-1 and 8-5).

Therefore, the relative displacement between the two helices will be $3\pi P/8\pi$, or $3P/8$, as shown in the original sketch by Watson and Crick (1953).

Since the intensities predicted by the above double helix on the first, second and third layer lines must locate at values of $2\pi rR$ correspond to the first peaks of $J_1^2(2\pi rR)$, $J_2^2(2\pi rR)$ and $J_3^2(2\pi rR)$, or roughly at 1.86, 3.05 and 4.20 respectively, and since the spot on the equatorial layer line corresponds to a value of $2\pi rR$ equal to about 3.14, we can conclude that the Watson and Crick double helical DNA secondary structure is in agreement with the X-ray diffraction picture of the lithium salt of DNA at 66% relative humidity (Fig. 8-3). In that picture, the intensity spot on the second layer line is located directly above that on the equatorial layer line, suggesting that the intensity on the second layer line indeed corresponds to $J_2^2(2\pi rR)$.

BIOLOGICAL CONSEQUENCES OF DNA DOUBLE HELIX

Indeed, the DNA double helical structure has been considered as the most fundamental aspect of modern day molecular biology. It consists of two essential parts: duality and helicity. Duality was also suggested by Chargaff's rule of [A] = [T] and [G] = [C]. Many molecular biological processes, e.g. DNA-DNA hybridization, semi-conservative replication, transcription, reverse transcription, DNA-RNA hybridization, etc., depend on this duality. On the other hand, helicity has resulted in numerous topological problems, as well as problems of energy requirement for unwinding and re-winding during DNA semi-conservative replication. One of the physicists turning molecular biologist once commented that in the highly viscous medium of the cell nucleus, our long DNA molecule of about 3.2 billion base-pairs had to unwind and re-wind during one cell cycles. The amount of energy lost due to viscous drag and then dissipated as heat would be very large.

To overcome the unwinding and re-winding problems, biochemists and molecular biologists have isolated numerous enzymes known as gyrases, helicases, topo-isomerases, etc. The scientific literature is filled with papers explaining how the problem of helicity can be resolved. Some of the topo-isomerases have been considered as "magicians' magician", since they can cut one of the DNA backbone strands, move it over to the other side of the other strand, and ligate the cut strand back without leaving a trace. This process will reduce the double helix by one turn. It is a fascinating story to read these articles. Many of the three-dimensional structures of these enzymes have been determined by X-ray crystallography.

Helicity causes another problem. Many of the biological active DNA molecules, e.g. that of plasmids, *E. coli*, etc., are double stranded intact circles. How can these molecules unwind and re-wind? Are cutting and pasting absolutely required? Why should evolution give rise to such DNA molecules with so many difficulties for them to replicate? Or, does evolution actually provide a vital advantage to such molecules? This has been an interesting problem to molecular biologists since 1953 when the DNA double helix was proposed.

All of such findings strongly suggest that we may need to look for another possible secondary structure of DNA. Interestingly enough, that structure was originally suggested by Wilkins, Stokes and Wilson in 1953 as:

"two similar coaxial helices of twice the above size and relatively displaced along the axis equal to half of the pitch".

Unfortunately, they discarded it by looking at the "width of layer line streaks", instead of the locations of the intensity spots on the various layer lines of the original picture obtained by Franklin and Gosling (1953) for sodium salt of DNA fibers at 92% relative humidity.

AN ALTERNATIVE HELICAL STRUCTURE OF DNA

For a single helix of pitch 2P, there will be twice the number of layer lines on the X-ray film representing the reciprocal space. Layer lines appear at $\zeta = n/2P$, instead of $\zeta = n/P$. In order to reduce the number of layer line by a factor of two, another helix of pitch 2P is placed along the same helical axis, but displaced by a distance P. With destructive interference, intensity spots on all odd layer lines will be cancelled, thus giving layer lines again at $\zeta = n/P$. As originally mentioned by Wilkins, Stokes and Wilson (1953), this combination of two helices will give spots on layer lines with intensities correspond to $J_2^2(2\pi rR)$, $J_4^2(2\pi rR)$, $J_6^2(2\pi rR)$, etc. Based on our previous discussion, the intensity spot on the first layer line will thus be located directly on top of that on the equatorial layer line. This is indeed the case for the X-ray diffraction picture shown in Fig. 8-1.

At the same time, the model must also result in reduced intensity on the first layer line and eliminate that on the fourth layer line. Thus, another similar set of two helices can be placed along the same helical axis and displaced relative to the first set by 3P/8 as discussed before (Figs. 8-4 and 8-5). The X-ray diffraction pattern based on this four-stranded structure, in its molecular details, is in perfect agreement with that shown in Fig. 8-1 (Wu, 1969). Another way of looking at that structure is to first consider one DNA double helix. Unwind it completely to a straight ladder with base-pairs on one side of the ladder. Take another ladder, and mutually intercalate the two ladders together. Then twist them down slightly. The mass per unit length remains the same.

In order to understand the intricacy of how various combinations of helices can give rise to different X-ray diffraction patterns, the following helices or combination of helices are illustrated:

1. one helix with radius r and pitch P (Fig. 8-8),
2. two co-axial helices of radius r and pitch P, relatively displaced for a distance of 3P/8 along the helical axis (Fig. 8-9),
3. one helix with radius r and pitch 2P (Fig. 8-10),
4. two co-axial helices of radius r and pitch 2P, relatively displaced for a distance of P along the helical axis (Fig. 8-11), and
5. two sets of two helices as in Fig. 8-11, relatively displaced for a distance of 3P/8 along the helical axis (Fig. 8-12).

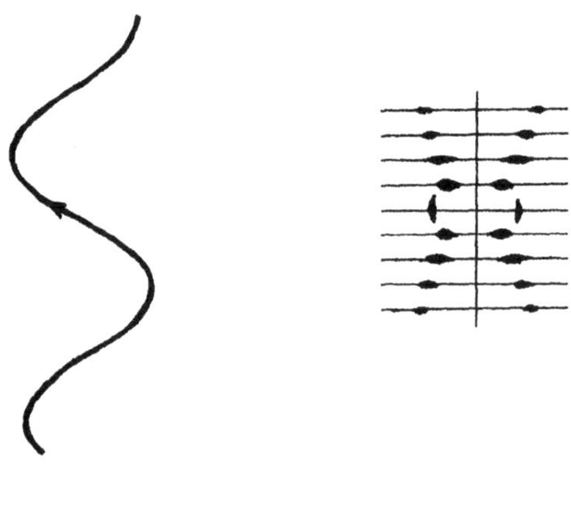

Figure 8-8. One helix with radius r and pitch P.

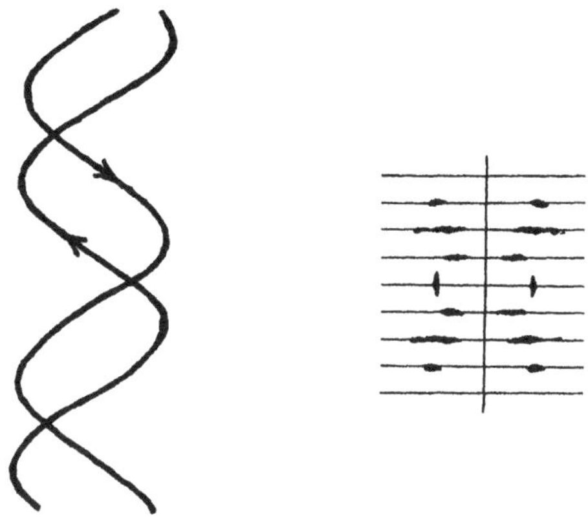

Figure 8-9. Two co-axial helices of radius r and pitch P, relatively displaced for a distance of 3P/8 along the helical axis.

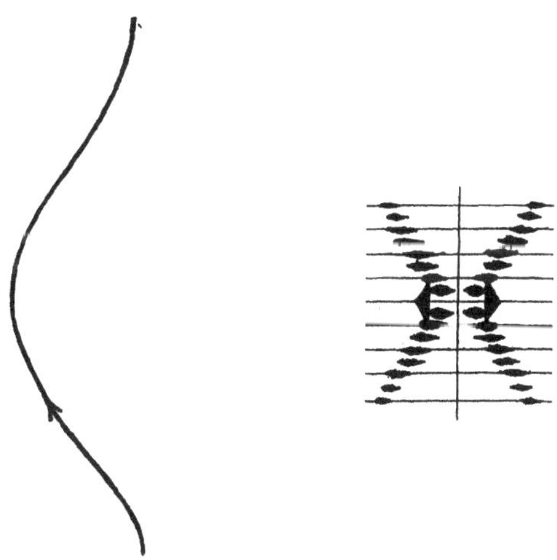

Figure 8-10. One helix with radius r and pitch 2P.

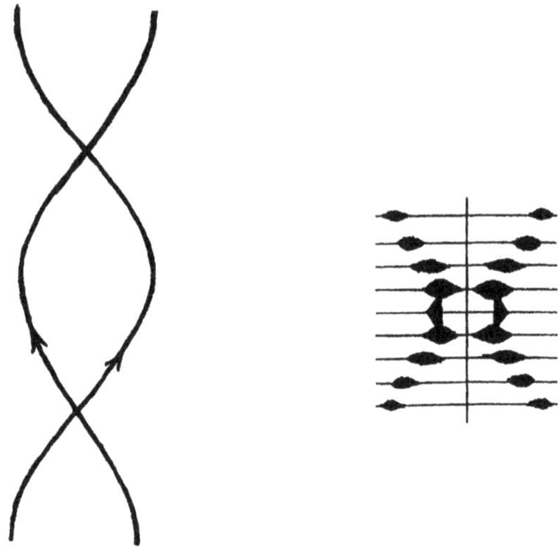

Figure 8-11. Two co-axial helices of radius r and pitch 2P, relatively displaced for a distance of P along the helical axis.

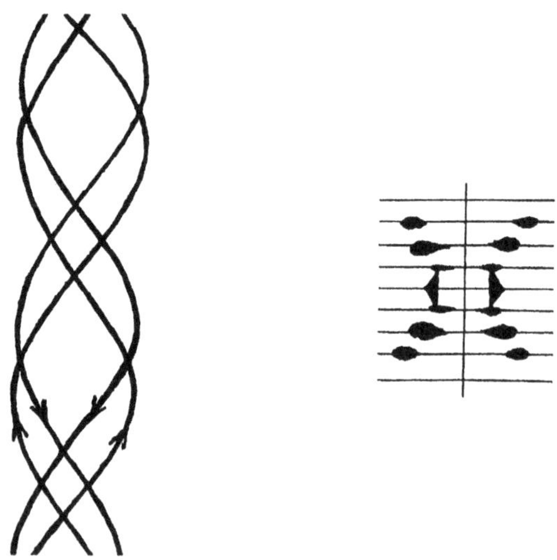

Figure 8-12. Two sets of two helices as in Figure 8-11, relatively displaced for a distance of 3P/8 along the helical axis.

In Figs. 8-8 to 8-12, the same scale is used for the reciprocal space coordinate system, **h**, and the spot on the equatorial layer line for R = 1/2r is shown. Then, it becomes obvious which combinations of helices give patterns similar to the experimental findings shown in Figs. 8-1 and 8-5. To study detailed molecular structures represented by these combinations will require massive amount of computing time. Numerical sums of various orders of Bessel function with double precision will be essential (Cochran, Crick and Vand, 1952).

Several other four-stranded helical structures of DNA have been proposed over the years. As pointed out by McGavin, Wilson and Barr (1966), there is no steric hindrance of the base-pairs in the middle of the helices in their model. Wilson is the same author for both this 1966 paper and the 1953 paper mentioned before. Similarly, Gehring, Leroy and Gueron (1993) have also proposed a sterically allowed four-stranded structure. Others consist of side-by-side double helices, some with one right-handed and another left-handed.

If DNA is double helical at 66% relative humidity and in crystals, and is quadruple helical at 92% relative humidity and in solution, a logical extrapolation to 100% relative humidity inside cell nucleus suggests that the sugar-phosphate backbone of DNA can exist in a straight configuration. This possibility can thus resolve many of the difficulties resulting from the universally accepted double helical structure, especially for unwinding and re-winding.

POSSIBLE BIOLOGICAL IMPLICATIONS

Possible biological implications of the above mathematical analysis of the X-ray diffraction pictures of DNA fibers will now be discussed.

Right-Handed or Left-Handed Helix

Do the X-ray diffraction pictures shown in Figs. 8-1 and 8-5 indicate that the DNA helix has to be right-handed?

For a physicist, he or she would place a mirror behind the experimental set-up and look at the mirror-image. There, the right-handed helix turn into left-handed helix, while the X-ray diffraction picture appears the same. That picture is symmetric right-to-left. Therefore, the physicist would conclude that if right-handed helix give that X-ray diffraction picture, left-handed helix would give the same picture. The solution is thus not unique.

For a mathematician, he or she starts with the equations of the right-handed helix or of the left-handed helix. There is only one negative sign difference. On carrying out the calculation for the total amplitude, **A(h)**, the mathematician discovers that $+\psi$ becomes $-\psi$. However, the intensity, **I(h)**, remains the same as $J_n^2(2\pi rR)$ since the phase drops out. Again, both the right-handed helix and left-handed helix give the same X-ray diffraction pattern.

For a biologist, he or she needs to learn some mathematics and physics as discussed here, before the above question can be answered. As we have pointed out, without a thorough knowledge of X-ray diffraction and the properties of Bessel function, it is rather difficulty to understand the entire problem.

Duality

All the consequences based on the duality of DNA double helix are retained for the four-stranded model.

Unwinding and re-winding

Since the sugar-phosphate backbones are straight, unwinding and re-winding are no longer required (Wu, 1969).

DNA Supercoils

The simplest DNA supercoils are the intact double-stranded DNA circles of plasmids. Under electron microscopes, they appear as rods or branched rods, each of which consists of two side-by-side DNA double helices according to the conventional explanation. The two complementary DNA circles are inter-twined so that they cannot separate from each other unless one of them is cut covalently. It is well known that many biological active DNA molecules need to retain their intact circular structure for replication. Therefore, in view of our present discussion, can this supercoil structure actually provide some evolutionary advantage?

We shall start with the four-stranded structure (Wu, 1969) as suggested by the X-ray diffraction picture of sodium salt of DNA fiber at 92% relative humidity, and consider it as two mutually intercalating stretched out double helices. We can connect their ends in such a way to give two base-paired circles of complementary strands. This rod structure (Wu and Wu, 1996; Stasiak, 1996) would be nearly indistinguishable from the conventional supercoil under the electron microscope. Nevertheless, in this case, the two complementary DNA single-stranded circles are not inter-twined and separable without any covalent bond breakage. Hopefully, with better resolutions of various types of electron microscopy techniques, the sugar-phosphate backbones of DNA supercoils may eventually be visible.

FUTURE EXPERIMENTS

As explained throughout this book, the interplay of experimentation by molecular biologists and analysis using mathematical and physical methods should suggest some future experiments. Since the current analysis of DNA secondary structure is highly unconventional and controversial, few or no experiments have been suggested and studied. Some will be suggested here.

Agarose Gel Electrophoresis

It is well known that intact plasmid supercoils moves faster than their linearized molecules on agarose gel electrophoresis. The usual explanation is that supercoils are more compact and thus move faster. However, one may ask the simple question of how much faster. Using linearized DNA molecular weight markers, one can easily measure the corresponding molecular weight of the intact supercoils. In most cases, they fall within about 15% of half of the molecular weight of the linearized molecules. This experiment should be verified with plasmids of different sizes. A possible explanation can consider the linearized molecules as representations of the conventional double helical DNA fiber structure. On the other hand, the intact supercoils may consist of two novel separable complementary single strand DNA circles, which are hydrogen-bonded to each other, representing the four-stranded DNA fiber structure suggested before. During agraose gel electrophoresis, they may separate due to the electrostatic repulsion of the negatively charged phosphate groups on their backbones. However, for most plasmid molecules, the molecular weights of the two complementary strands are very close to each other. Thus, even if they have separated, the two complementary single stranded DNA circles would appear as a single band. New experimental methods would be required to study how they can be distinguished.

Effect of mRNA

Many of the plasmid molecules carry drug resistant markers, i.e. they direct the synthesis of enzymes which can degrade certain antibiotics. Thus, if the plasmid molecules are purified under conditions so that the mRNA molecules remain intact, these mRNA molecules will bind to one of the two complementary strands of DNA selectively. Since RNA-DNA hybridization is more stable than DNA-DNA hybridization, it will be possible to select conditions for agarose gel electrophoresis such that the mRNA molecules are still associated with one of the DNA strands. The molecular weight of that DNA strand together with the mRNA is thus larger than the other DNA strand alone. Furthermore, the configuration of the former may also be different.

Experimentally, the two complementary intact single strand DNA circles of plasmid molecules, in the presence of mRNA molecules binding selectively to one of them, can separate into two bands on agarose gel electrophoresis (Wu and Wu, 1996). Each of the bands can then be concentrated with the glass bead method, and their nucleotide sequences determined. For pUC19, one of the smaller plasmids, their sequences are indeed complementary to each other. This experiment must be repeated for other plasmid molecules of larger molecular weights, in order to establish that the complementary strands of all plasmids are indeed separable regardless of their lengths. Agarose gel electrophoresis may have a length limit of about 50 kb for such studies, as in the case of cosmid molecules.

Programmed Electrophoresis

Even though the two complementary strands of most plasmids have very similar molecular weights, they may have different folding configurations when separated from each other. Therefore, after running agarose gel electrophoresis in one direction for a certain period of time, we can reverse the electrodes and run it in the other direction. The duration of each run can vary by changing the current or voltage. In addition, the condition of the buffer can also be adjusted. The configuration of one of the two complementary circles may then move faster. Thus, current, time, pH, direction, and other factors of the run can be altered in order to get the best separation of the two bands. Depending on the physical length of the agarose gel, we can have a programmed electrophoresis for plasmid molecules of different molecular weights. If this idea works, the separated bands can again be concentrated and analyzed, or be hybridized with various probes.

Restriction Enzyme Activities

It has been noted that most restriction enzymes can digest linearized DNA molecules faster than DNA supercoils. A careful quantitative analysis of these enzymatic activity may support the idea that linearized DNA molecules are in the conventional double helical configuration and that restriction enzymes usually bind to double helices. On the other hand, DNA supercoils can exist in the four-stranded structure as discussed before, and may not bind these enzymes effectively. However, local conversion from one structure to the other in equilibrium can give rise to reduced rates of digestion.

Binding of Topo-isomerase I

During recent years, topo-isomerase I is commercially available. Presumably, in the near future, other topo-isomerases, as well as gyrases, helicases, etc. will also be available to the molecular biology community. Their interactions with DNA supercoils and linearized DNA molecules can then be investigated carefully. For example, in the presence of mRNA, intact pUC19 molecules give two bands on agarose gel electrophoresis.

However, with the addition of topo-isomerase I, the two bands collapse into one. Furthermore, under that condition, digestion by restriction enzymes is even more retarded. How are these biological macro-molecules interacting with each other (see, for example, Arimondo *et al.*, 2000)?

Three Dimensional Folding of DNA

Since the discovery of DNA double helix more than 40 years ago, the three-dimensional folding of DNA, or DNA tertiary structure, is largely unknown. How is the DNA molecule packed inside the head of a lambda phage? On osmotic shock, the phage DNA will exhibit many loops. Since the detailed folding of this 50 kb DNA molecule is not understood, larger DNA molecules are essentially beyond comprehension. Various basic proteins are intimately associated with DNA, but the detailed protein-DNA interactions are also not understood, except in a few cases. As discussed in the previous chapter, the study of three-dimensional structures of DNA will be a major endeavor in the future.

REFERENCES

Abramowitz M and Stegun IA (1965) *Handbook of Mathematical Functions.* National Bureau of Standards, Applied Mathematics Series 55. Washington, DC: U.S. Government Printing Office.

Arimondo PB, Riou J-F, Mergny J-L, Tazi J, Sun J-S, Garestier T and Helene C (2000) Interaction of human DNA topoisomerase I with G-quartet structures. *Nucl. Acids Res.*, **28**, 4832-4838.

Cochran W, Crick FHC and Vand V (1952) The structure of synthetic polypeptides. I. The transforms of atoms on a helix. *Acta Cryst.*, **5**, 581-586.

Franklin RE and Gosling RG (1953) Molecular configuration in sodium thymonucleate. *Nature,* **171**, 740-741.

Gehring K, Leroy J-L and Gueron M (1993) A tetrameric DNA structure with protonated cytosine-cytosine base pairs. *Nature,* **363**, 561-565.

Langridge R, Wilson HR, Cooper CW, Wilkins MHF and Hamilton LD (1960) The molecular configuration of deoxyribonucleic acid. I. X-ray diffraction study of crystalline form of lithium salt. *J. Mol. Biol.*, **2**, 19-37.

McGavin S, Wilson HR and Barr GC (1966) Intercalated nucleic acid double helices: a stereochemical possibility. *J. Mol. Biol.*, **22**, 187-191.

Stasiak A (1996) Getting down to the core of homologous recombination. *Science,* **272**,828-829.

Watson JD and Crick FHC (1953) Molecular structure of nucleic acids, *Nature,* **171**, 737-738.

Whittaker ET and Watson GN (1915) *A Course of Modern Analysis.* Cambridge: at the University Press.

Wilkins MHF, Stokes AR and Wilson HR (1953) Molecular structure of deoxypentose nucleic acids. *Nature,* **171**, 738-740.

Wu TT (1969) Secondary structure of DNA. *Proc. Nat. Acad. Sci. USA,* **63**, 400-405.

Wu R and Wu TT (1996) A novel intact circular dsDNA supercoil. *Bull. Math. Biol.*, **58**, 1171-1185.

EXERCISE

Determine the X-ray diffraction pattern of three identical co-axial helices each with radius r and pitch 3P and relatively displaced from each other by P along the helical axis.

APPENDIX

INTRODUCTION

During my numerous discussions with my colleagues in various departments in different universities about writing this book, I have been reminded repeatedly to include an appendix to review various simple mathematical methods for the benefit of readers who may have forgotten about them. I also remember an incident many years ago about a first year graduate student from my Department of Biochemistry, Molecular Biology and Cell Biology. She wanted to spend three months in my laboratory so that she could have a better understanding of the Hill's plot. I asked her whether she still remembered the definition of logarithm. She looked at me and said, "Of course!" I was a little worried, and said gently, "Can you tell me what it is?" She replied, "It is a button on my calculator."

We have learned many mathematical methods during our high school days. Unfortunately, we have also forgotten most of them. In addition, other mathematical methods are covered in various courses in college. Unless they are applied immediately to solve problems, they are also forgotten. So, to begin with, the definition of logarithm will be discussed.

LOGARITHM

For numbers like thousand, million, billion or trillion, they can be represented as 10^3, 10^6, 10^9 and 10^{12}. Logarithm with base 10 is then defined as, for example:

$$\log_{10} 10^3 = 3 .$$

In general, we have:

$$\log_{10} x^n = n \log_{10} x .$$

The number 10 is arbitrary, and we can have other numbers. However, there is a number denoted by e and defined as:

$$e = \lim_{n \to \infty} (1 + 1/n)^n .$$

Logarithm with base e (= 2.71829) is usually denoted by ln. Similarly, we have:

$$\ln x^n = n \ln x .$$

The usefulness of the number e will become clear when we discuss about derivatives.

In Hill's plot, ln is used so that the non-linear equation based on the Hill's model of hemoglobin saturation with oxygen at equilibrium can be converted into a straight line (see Chapter 2).

SIMULTANEOUS LINEAR ALGEBRAIC EQUATIONS

For three unknowns related by linear algebraic equations, they can be solved using determinants. Consider the following set of equations:

$$ax + by + cz = A,$$

$$ex + fy + gz = B,$$

and

$$kx + my + nz = C,$$

where x, y and z are the unknowns, while the other symbols represent known quantities. Then, the value of x is given by the following equation:

$$x = \frac{\begin{vmatrix} A & b & c \\ B & f & g \\ C & m & n \end{vmatrix}}{\begin{vmatrix} a & b & c \\ e & f & g \\ k & m & n \end{vmatrix}}.$$

Values of y and z can be similarly expressed. The value of the determinant is calculated as follows:

$$\begin{vmatrix} a & b & c \\ e & f & g \\ k & m & n \end{vmatrix} = a(fn - mg) - b(en - kg) + c(em - kf).$$

This result is used in Chapter 4 to study non-competitive enzyme inhibitors.

QUADRATIC ALGEBRAIC EQUATION

Quadratic equations appear frequently in many mathematical problems. The two solutions can be obtained using some simple algebraic relations. Consider:

$$A\,x^2 + B\,x + C = 0\,,$$

where A, B and C are constants. Dividing the equation by A gives:

$$x^2 + (B/A)\,x + (C/A) = 0\,.$$

Then, add on a term so that the first two terms can be combined to form a perfect square, i.e.

$$x^2 + (B/A)\,x + (B/2A)^2 - (B/2A)^2 + (C/A) = 0\,,$$

or
$$\{x + (B/2A)\}^2 - (B^2 - 4AC)/(2A)^2 = 0\,,$$

or
$$\{x + (B/2A) - (B^2 - 4AC)^{1/2}/2A\}\cdot\{x + (B/2A) + (B^2 - 4AC)^{1/2}/2A\} = 0\,.$$

Therefore, we get:

$$x = \{-B + (B^2 - 4AC)^{1/2}\}/2A\,,$$

or
$$x = \{-B - (B^2 - 4AC)^{1/2}\}/2A\,.$$

This result is used to solve for α^3 in the Pauling's model of hemoglobin saturation (see Chapter 2).

TRANSLATION AND ROTATION OF CARTESIAN COORDINATES

In two dimensions, to translate the original of one Cartesian coordinate system (x,y) to a point (a,b) results in a new coordinate system (x',y') as shown in Fig. A-1.

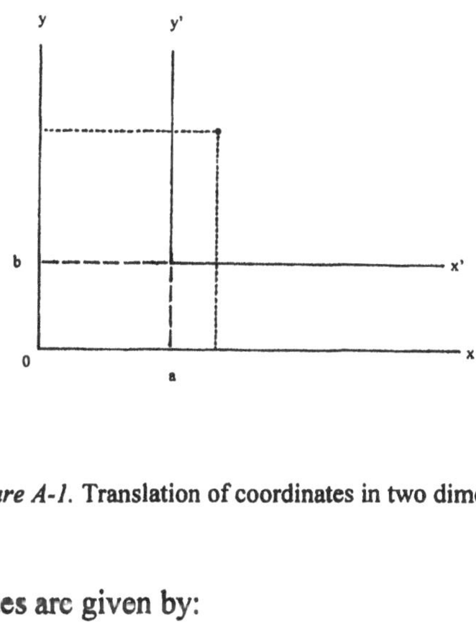

Figure A-1. Translation of coordinates in two dimension.

The new coordinates are given by:

$$x' = x - a,$$

and

$$y' = y - b.$$

In two-dimensional rotation, let the angle of rotation at the origin by θ. As shown in Fig. A-2, a point represented by (x,y) in the original coordinate system is then denoted as (x',y') in the new coordinate system. The values of x' and y' can be calculates by the construction illustrated in that figure.

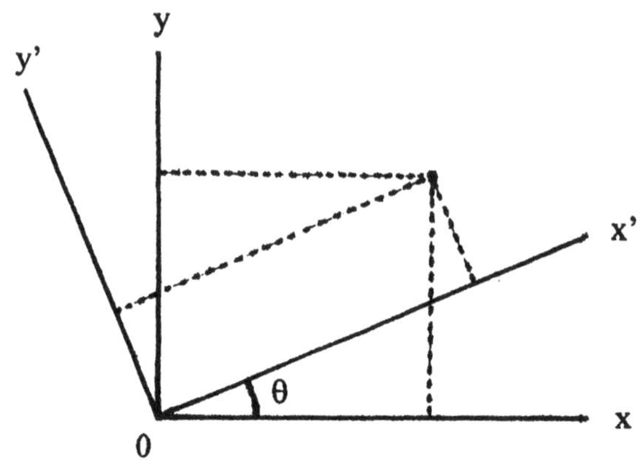

Figure A-2. Rotation of coordinates in two dimension.

Using the definitions of sine and cosine, we get:

$$x' = \cos\theta \, x + \sin\theta \, y,$$

and

$$y' = -\sin\theta \, x + \cos\theta \, y.$$

In matrix notation, the last two equations can be written as:

$$\begin{bmatrix} x' \\ y' \end{bmatrix} = \begin{bmatrix} \cos\theta & \sin\theta \\ -\sin\theta & \cos\theta \end{bmatrix} \begin{bmatrix} x \\ y \end{bmatrix}.$$

The 2 X 2 matrix on the right-hand side is usually referred to as the rotation matrix.

In three dimensions, translation is similar, i.e.:

$$x' = x - a,$$

$$y' = y - b,$$

and $$z' = z - c.$$

Rotation is a combination of three two-dimensional rotations, i.e. obtaining the coordinates by rotating around three axes separately. These angles of rotation are referred to as Eulerian angles. Unfortunately, there is no universally agreed convention of the sequence and direction of rotations. One possible set of Eulerian angles, ϕ, θ and ψ, gives the following rotation matrix in three-dimension (Goldstein, 1950):

$$\begin{bmatrix} \cos\psi\cos\phi - \cos\theta\sin\phi\sin\psi & \cos\psi\sin\phi + \cos\theta\cos\phi\sin\psi & \sin\psi\sin\theta \\ -\sin\psi\cos\phi - \cos\theta\sin\phi\sin\psi & -\sin\psi\sin\phi + \cos\theta\cos\phi\sin\psi & \cos\psi\sin\theta \\ \sin\theta\sin\phi & -\sin\theta\cos\phi & \cos\theta \end{bmatrix}$$

However, for the discussion of rotations for single bonds in polypeptides and polynucleotides (Chapters 5 to 7), successive two-dimensional rotations can be chosen to simplify calculations of atomic coordinates.

DERIVATIVE

If y is a function of x, we usually write:

$$y = y(x).$$

If y is plotted against x, we can think about the function as a curve. The slope of the curve at any given value of x is defined as:

$$\frac{dy}{dx} = \lim_{\Delta x \to 0} \frac{y(x + \Delta x) - y(x)}{\Delta x}.$$

It can be easily verified that:

$$dx / dx = 1,$$

$$dx^2 / dx = 2x,$$

$$dx^3 / dx = 3x^2,$$

$$dx^n / dx = n x^{n-1}.$$

by power series expansion, e,g,

$$(x + \Delta x)^2 = x^2 + 2 x \Delta x + (\Delta x)^2,$$

and $(x + \Delta x)^n = x^n + n x^{n-1} +$ higher order terms in Δx.

Using the definition of e, we can calculate the derivative of e^x with respect to x:

$$de^x / dx = d\{ \lim_{n \to \infty} (1 + 1/n)^{nx}\} / dx$$

$$= \lim_{\Delta x \to 0} \lim_{n \to \infty} \{(1 + 1/n)^{n(x+\Delta x)} - (1 + 1/n)^{nx}\} / \Delta x$$

$$= \lim_{n \to \infty} (1 + 1/n)^{nx} \lim_{\Delta x \to 0} \{(1 + 1/n)^{n\Delta x} - 1\} / \Delta x$$

$$= \lim_{n \to \infty} (1 + 1/n)^{nx} \lim_{\Delta x \to 0} \Delta x / \Delta x$$

$$= e^x .$$

This expression is sometimes also denoted as exp[x]. For this reason, ln is known as the natural logarithm.

Partial derivative for functions of two variables with respect to one of the variables can be obtained by considering the other variable as a constant. For example,

$$\frac{\exp[x(t-1/t)/2]}{x} = (t-1/t)/2 \exp[x(t-1/t)/2] ,$$

as discussed in Chapter 8.

INTEGRATION

For simplicity, integration is considered as the reverse of derivative. Thus:

$$\int x^n \, dx = (n+1)^{-1} x^{n+1}.$$

This result is used in Chapter 3. Similarly for terms involving e, we have:

$$\int e^{nx} \, dx = n^{-1} e^{nx},$$

which is used in Chapter 4.

For definite integrals with limits of integration, the value is calculated for the upper limit and is subtracted with that calculated at the lower limit.

VECTORS

For two vectors in three dimension:

$$\mathbf{a} = a_x\,\mathbf{i} + a_y\,\mathbf{j} + a_z\,\mathbf{k},$$

and

$$\mathbf{b} = b_x\,\mathbf{i} + b_y\,\mathbf{j} + b_z\,\mathbf{k},$$

where **i**, **j** and **k** are unit vectors along the x-, y- and z-axis. Their dot product gives a scalar:

$$\mathbf{a}\cdot\mathbf{b} = a_x\,b_x + a_y\,b_y + a_z\,b_z\,.$$

Their cross product is defined as:

$$\mathbf{a}\,\mathrm{X}\,\mathbf{b} = \begin{vmatrix} \mathbf{i} & \mathbf{j} & \mathbf{k} \\ a_x & a_y & a_z \\ b_x & b_y & b_z \end{vmatrix}$$

$$= (a_y\,b_z - a_z\,b_y)\,\mathbf{i}$$

$$+ (a_z\,b_x - a_x\,b_z)\,\mathbf{j}$$

$$+ (a_x\,b_y - a_y\,b_x)\,\mathbf{k}$$

The direction of this vector is perpendicular to both **a** and **b** and follows the right-hand rule.

The length of vector **a** equals to:

$$|\mathbf{a}| = \{a_x^2 + a_y^2 + a_z^2\}^{1/2}.$$

Then, we also have for the dot product:

$$\mathbf{a} \cdot \mathbf{b} = |\mathbf{a}|\ |\mathbf{b}| \cos\theta ,$$

where θ is the angle between vectors **a** and **b**. Similarly, for the cross product:

$$|\mathbf{a} \times \mathbf{b}| = |\mathbf{a}|\ |\mathbf{b}| \sin\theta .$$

The tetrahedral angle, τ, can be calculated with the use of the dot product. Consider

$$\mathbf{a} = \mathbf{i} + \mathbf{j} + \mathbf{k} ,$$

and

$$\mathbf{b} = -\mathbf{i} - \mathbf{j} + \mathbf{k} .$$

Therefore,

$$-1 = 3 \cos\tau ,$$

or

$$\cos\tau = -1/3 ,$$

i.e.

$$\tau = 109.47^{\circ} ,$$

which is usually rounded to 110° (Chapter 5).

SIMPLE DIFFERENTIAL EQUATION

From the definition of e, the solution of the following differential equation:

$$dy/dt = Ay$$

is simply $y = Be^{At}$, where A and B are constants. This equation is commonly encounter in many biological processes. For example, a culture of bacteria grows exponentially can be described for A being positive. If the starting number of bacteria is N_0, then $B = N_0$. The doubling time is calculated from:

$$2 = e^{At},$$

or

$$t = (\ln 2)/A.$$

For any decaying process, whether a radioactive material, HIV-1 RNA molecules in blood of AIDs patients, or other substances, A is negative. In such cases, a similar calculation gives the half-life.

The above differential equation is classified as linear first order ordinary. Other differential equations can be of higher order, non-linear, partial, etc. (see, for example, Taubes, 2001). They can also be converted into integral equations.

COMPLEX VARIABLE

The complex variable is usually represented by $z = x + i\,y$, where $i = (-1)^{1/2}$. A function of complex variable, f(z), is said to be regular or analytic in a region of the complex plane, if its derivative exists (see, for example, Carrier *et al.*, 1966). The complex plane has a horizontal or real axis denoted by x, and a vertical or imaginary axis denoted by y. The most important theorem of complex variable is the Cauchy integral theorem (Knopp, 1943):

$$\oint f(z)\,dz = 0\,,$$

for a contour which encloses a region where f(z) is analytic. The arrow indicates that the integration is in the counter-clockwise direction (Fig. A-3).

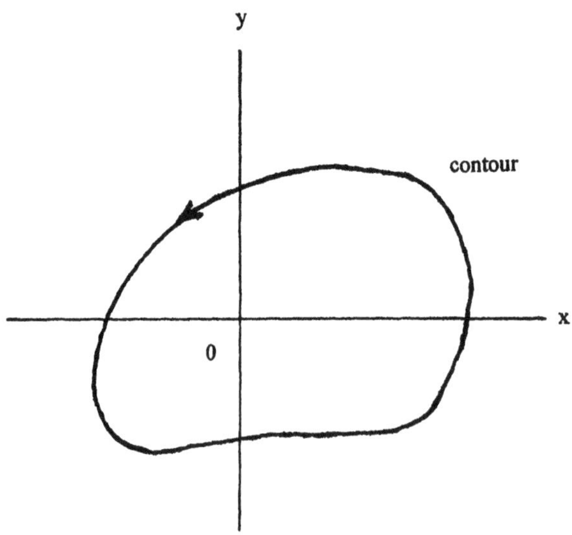

Figure A-3. Contour integral in the complex plane.

If the contour is a unit circle around the origin of the complex plane, then according to the Cauchy integral theorem we have:

$$\oint z^n \, dz = 0 ,$$

for integer n equal to or greater than 0. For n = –1, we substitute z with $e^{i\theta}$ so that $dz = i\, e^{i\theta}\, d\theta$. Then,

$$\oint z^{-1} \, dz = \int_{\alpha}^{\alpha+2\pi} i \, d\theta = 2\pi i ,$$

where α is any starting angle for integration. For n = –2, the same change of variable gives:

$$\oint z^{-2} \, dz = \int_{\alpha}^{\alpha+2\pi} i\, e^{-i\theta} \, d\theta = -i\, e^{-i\theta} \Big|_{\alpha}^{\alpha+2\pi} = 0 .$$

The same holds for n = –3, –4, –5, etc. These results are used in Chapter 8 to calculate the integral representation of the Bessel functions.

FOURIER TRANSFORM

As we have discussed in Chapter 8, X-ray diffraction study connects the physical space **x** to the reciprocal space **h**. Similarly, time t can be related to frequency λ. Such connections are through a mathematical operation known as Fourier transform. The Fourier transform F(λ) of a function f(t) can be defined as (Carrier *et al.*, 1966):

$$F(\lambda) = (2\pi)^{-1/2} \int_{-\infty}^{\infty} e^{i\lambda t} f(t)\, dt ,$$

and the inverse Fourier transform is:

$$f(t) = (2\pi)^{-1/2} \int_{-\infty}^{\infty} e^{-i\lambda t} F(\lambda)\, d\lambda .$$

This transform and many other transforms are valuable mathematical tools to solve differential and integral equations. One of the important results is the Convolution Integral which gives an expression for the inverse Fourier transform of the product of two Fourier transforms:

$$(2\pi)^{-1/2} \int_{-\infty}^{\infty} e^{-i\lambda t} F(\lambda)\, G(\lambda)\, d\lambda = (2\pi)^{-1/2} \int_{-\infty}^{\infty} f(x - \xi)\, g(\xi)\, d\xi .$$

Cochran *et al.* (1952) used this result to derive the X-ray diffraction pattern of a discontinuous helix as the summation of various Bessel functions.

REFERENCES

Carrier GF, Krook M and Pearson CE (1966) *Functions of a Complex Variable*. McGraw-Hill Book Company, New York.

Cochran W, Crick FHC and Vand V (1952) The structure of synthetic polypeptides. I. The transform of atoms on a helix. *Acta Cryst.*, 5, 581-586.

Goldstein H (1950) *Classical Mechanics*. Addison-Wesley Publishing Company, Inc., Reading, MA.

Knopp K (1943) *Theory of Functions*. Dover Publications, Inc., New York.

Taubes CH (2001) *Modeling Differential Equations in Biology*. Prentice Hall, Upper Saddle River, NJ.

AUTHOR INDEX

SUBJECT INDEX

MIX
Papier aus verantwortungsvollen Quellen
Paper from responsible sources
FSC® C105338

If you have any concerns about our products,
you can contact us on
ProductSafety@springernature.com

In case Publisher is established outside the EU,
the EU authorized representative is:
Springer Nature Customer Service Center GmbH
Europaplatz 3, 69115 Heidelberg, Germany

Printed by Libri Plureos GmbH
in Hamburg, Germany